PRINCIPLES OF MACHINE OPERATION AND MAINTENANCE

PRINCIPLES OF MACHINE OPERATION AND MAINTENANCE

Dick Jeffrey

Routledge
Taylor & Francis Group

LONDON AND NEW YORK

First published by Butterworth-Heinemann

This edition published 2011 by Routledge
2 Park Square, Milton Park, Abingdon, Oxon OX14 4RN
711 Third Avenue, New York, NY 10017, USA

Routledge is an imprint of the Taylor & Francis Group, an informa business

First edition published by Thomas Nelson Australia 1991
First Published in Great Britain 1991

Notice
No responsibility is assumed by the publisher for any injury and/or damage to persons
or property as a matter of products liability, negligence or otherwise, or from any use
or operation of any methods, products, instructions or ideas contained in the material
herein. Because of rapid advances in the medical sciences, in particular, independent
verification of diagnoses and drug dosages should be made

British Library Cataloguing in Publication Data
A catalogue record for this book is available from the British Library

Library of Congress Cataloging-in-Publication Data
A catalog record for this book is available from the Library of Congress

ISBN: 978-0-7506-0293-8

CONTENTS

Preface

Principles of Machine Operation and Maintenance has been designed both as a training manual for maintenance tradespeople and technicians and as a reference handbook for engineers. It sets out to achieve three things:
- to establish and describe the basic principles of operation and maintenance of rotating machinery,
- to create a clear understanding of the maintenance function and the role of the technician,
- to emphasise the importance of a positive mental approach to maintenance and the need to develop troubleshooting skills as well as practical skills.

The book assumes that the reader will have already acquired a basic understanding of engineering terms and practices such as would be gained during the first stages of a fitting and machining apprenticeship.

The illustrations provided have been kept deliberately simple and schematic in nature so that basic principles rather than the precise details of machine element construction can be emphasised.

The underlying aim of the book has been to promote the principles of good engineering practice and it is hoped that it will provide a basis of understanding on which experience can be assimilated.

Thanks are extended to Mobil Oil Australia Ltd, SKF Australia Ltd, Glacier Metal Company Ltd, the Gates Rubber Company, Denver, Colorado, Rex Chainbelt Inc, Milwaukee, Caterpillar of Australia Limited, Renold Australia P/L, Wetherby Engineering Co. Ltd (Broadbent Drives), Bradford, UK, Stieber Ltd, Letchworth, UK, and Flexibox P/L for the use of illustrative material. I should also like to thank Ruth Siems for her expert editing and her sympathetic and patient treatment of a novice writer and Sandra McComb for her encouragement and faith in the project.

Dick Jeffrey
October 1984

C H A P T E R 1

THE MAINTENANCE FUNCTION

The function of maintenance is to ensure that plant and equipment are available in a satisfactory condition for operation when required. The determination of what constitutes a 'satisfactory condition' for rotating machinery will depend largely on the operating situation and the following considerations:

Type of industry The nature of the industry concerned will determine, to some extent, the level of machine performance required.

Process requirements The function of the machine in the process will also determine performance requirements. The requirements of a machine that is critical to the process will be greater than for one that is 'non-critical'.

Business objectives Product demand and profit levels may give rise to fluctuating demands on equipment and performance requirements.

In all cases, however, the performance of the maintenance function can be judged by the condition of machinery as indicated by the following factors:

Performance Machines must be capable of performing the function for which they are intended.

Downtime Machines must operate with an acceptable level of downtime.

Service life Machines must provide a satisfactory return on investment before replacement becomes necessary.

Efficiency Machines must operate at an acceptable level of efficiency.

Safety Machines must operate safely and not be dangerous to personnel.

Environmental impact Machines must operate in a manner that is not detrimental to the environment or to adjacent plant and equipment.

Cost The cost of maintenance must be acceptable.

The goal of maintenance is to ensure that machinery performance is 'satisfactory' according to these seven factors and, although the specific requirements of an individual machine are rarely quantified, it is important that the criteria by which performance can be assessed are understood and monitored. Despite the fact that definite levels of acceptibility are hard to establish, trends in machine conditions can be observed and should be used as indicators of maintenance requirements.

PERFORMING THE MAINTENANCE FUNCTION

The maintenance function is performed in two ways:
- by the prevention of breakdown, and
- by the repair of breakdown.

The term, breakdown, is used here to indicate any machine condition that is considered to be less than satisfactory according to the seven factors listed above. This covers all situations from the need for minor adjustment to total machine collapse.

Although performance of the maintenance function has been separated here into 'prevention' and 'repair', the element of prevention is the key to all successful maintenance work, including repair. Preventative maintenance need not be treated as a special program because prevention should be the underlying theme of all maintenance work.

Rather than investigating repair and prevention as two different aspects, it is more useful to consider the various ways in which maintenance work arises and to recognise the preventative element in each.

Machine breakdown

Despite all attempts at prevention, machine breakdowns of various kinds do occur and often need to be fixed on an urgent or emergency basis. Although maintenance personnel are usually under great pressure to return equipment to service in the shortest possible time, there are two important considerations, vital to the 'preventative' element of the work, that should not be sacrificed.

Treatment of cause — not effect

It is important to make sure that the real cause of the breakdown is found and remedied and not just the effect patched up. Troubleshooting should be rigorous in finding the inherent cause of the problem. If the real cause of the problem is not corrected then further breakdown is likely to occur.

Preventative solutions

Restoration of equipment to its original condition is not necessarily the best solution to a breakdown problem. Consideration should be given to avoiding repetition of the problem by making minor changes in design or materials. Equipment performance can often be improved with little expense by the development of innovative solutions that are specially suited to local conditions and process requirements.

Routine work

Some basic preventative work can be easily identified for execution on a routine basis. This work will typically include such tasks as:
  lubrication
  adjustment
  cleaning
and the need for such work will usually be determined on the basis of:
  field experience
  manufacturer's recommendations
  production requirements
Routine work should, where possible, be carried out during machine operation. If this is not possible it may be scheduled to coincide with regular production shutdowns. In any event, this type of work should not normally interfere with production schedules.

Planned work

This category includes any work that requires an extended equipment shutdown or that needs to be planned and scheduled in advance.

All planned work should be considered as preventative and is essentially aimed at the avoidance of unscheduled breakdowns. The type of work involved will include:
  major overhaul
  replacement of specific components
  machine modification
It may originate from inspection, condition monitoring, manufacturer's recommendations, plant experience, statutory regulations or design modification.

Inspection

The condition of all machinery should be under continual surveillance by both operating and maintenance personnel. The casual and routine monitoring of equipment will yield information regarding operating condition on which maintenance requirements can be planned.

It is vital that maintenance personnel realise the importance of being critically aware of the operating condition of machinery and ensure that their observations are accurately reported. Inspection requires the use of the senses and maintenance personnel should develop an **eye** and an **ear**, and even a **nose**, for machine condition. Recognition of normal running characteristics are the basis from which deviations can be observed and trends in machine condition can be predicted.

Condition monitoring

In recent years a variety of techniques have been developed by which the operating condition of machinery can be either intermittently or continuously monitored. These techniques are a mechanised version of inspection by personnel and all operate on the same principle of observed deviation from normal condition.

The most important of these techniques is **vibration monitoring** and this and other methods are outlined in Chapter 10. Condition monitoring, like personal inspection, yields information on which maintenance requirements can be based.

Manufacturers' recommendations

Most equipment manufacturers provide details of recommended maintenance requirements, from basic lubrication schedules to major overhaul information. Until plant experience indicates otherwise, it is wise to follow these recommendations in the early stages of operation. This information provides an initial basis on which to determine preventative work, such as overhaul and routine replacement of components, to be carried out during annual or other planned shutdowns.

Plant experience

Local knowledge of machine performance under specific plant conditions may well, in the longer term, prove to be a more appropriate method of establishing the frequency of major overhaul and other maintenance requirements. For such experience to be used effectively it is important that accurate maintenance records are kept so that performance patterns and characteristics can be clearly established.

Component life expectancy and wear rates can only be assessed on the basis of recorded information that represents a true reflection of operating conditions. The use of condition monitoring techniques provides the kind of detailed information on which maintenance records can be based.

Statutory regulations

Inspection and testing of certain types of equipment, especially pressure vessels, is normally carried out on an annual basis according to statutory regulations. In most cases this involves a total plant shutdown and is an opportunity to undertake an annual 'turn around'. A certain amount of work is generated during this period due to the inspection program itself as determined by the regulations, and the remainder originates in the ways previously described. Compliance with statutory regulations constitutes, in effect, an externally imposed prevention program.

Design modifications

When the performance of a machine is unsatisfactory according to any of the seven factors listed on page 1, then design modification may be undertaken in order to improve performance.

Modifications will normally be undertaken by engineering staff who support maintenance personnel. Unless the machine has already suffered major breakdown, in which case modification may be undertaken on an emergency basis, the work involved would normally be scheduled for a planned shutdown period.

No matter what the source of the work, it should be clear from the above descriptions that the common element is the prevention of future breakdown. Although no maintenance department can ever be one hundred per cent successful in this, it should be recognised that the general objective of **breakdown prevention** helps to encourage an attitude towards maintenance work that is most likely to give good results.

TRADE SKILLS

Whatever the philosophical approach and organisational methods adopted, it is the maintenance technician, the person who actually performs the work, who is the focal point of the maintenance operation. The technician constitutes the key element in the operation around whom other resources must be organised. In order to fulfil this critical function technicians must possess certain skills, both practical and theoretical.

Practical skills

A maintenance technician should have completed an apprenticeship in fitting and machining (or the local equivalent). Where elective components are available in the apprenticeship program, options that concentrate on fitting, and particularly maintenance fitting, should obviously be selected where possible. This training should provide all the basic practical skills needed to undertake maintenance work. Such skills will include all basic manipulative tasks associated with fitting and machining including the use of appropriate tools and equipment. The level of competence and the range of skills involved will normally be determined by the relevant regulatory body (Industrial Training Commission or otherwise).

The skills taught in an apprenticeship program represent the minimum requirement and the practical skills of maintenance technicians should be further developed on the job to meet specific plant requirements and to keep pace with technological development.

Theoretical skills

In addition to the theoretical background provided to complement the practical skills acquired during an apprenticeship, technicians should possess an understanding of the operating principles of basic machine elements and common types of machinery as well as the principles and practices of maintenance associated with them. The theory component of an apprenticeship is normally too basic and general in nature to provide an adequate background for maintenance work. In order to be fully equipped to troubleshoot and to make effective repairs, the technician must have a clear understanding of the function and principles of operation of the machinery involved.

These theoretical skills, specifically related to rotating machinery, are the primary focus of this book.

Mental attitude

Although it is often neglected, one of the most important skills that a technician can possess is a good mental attitude towards the maintenance task. The most obvious example of this is the way in which a careless and casual approach is likely to lead to error and bad workmanship. The adoption of a positive mental attitude toward the task not only has the potential to improve performance but also leads to increased personal satisfaction. Some of the ways in which mental attitude can be improved are discussed in Chapter 12.

Troubleshooting skills

A vital skill in all aspects of maintenance work that assumes even greater importance when the technician takes on a supervisory role is the ability to troubleshoot and diagnose faults. Like most skills, troubleshooting ability can be learned. It is, in a sense, the key element in the technician's resources drawing together practical and theoretical skills to produce the most effective performance.

Troubleshooting skills are largely an extension of certain aspects of mental attitude and are discussed in detail in Chapter 11.

RESOURCES

To complement the technician's personal skills, there is a range of additional resources that should be available within the maintenance department. The responsibility for organising these resources so that they are available when and where required lies with management, but it is the responsibility of the technician to appreciate their nature and to understand how they can be most effectively utilised.

The exact nature of these resources will vary from one organisation to another but the following list identifies the principal resources that should be available.

Tools	Standard hand tools
	Specialised tools
	Workshop facilities
Materials	Spare parts
	Stock materials
	Standard consumables
	Salvaged stock
Expertise	Line supervision
	Staff engineers
	Consultants
Systems data	Plant records
	Engineering drawings
	Process flow diagrams
	Manufacturers' information
	Technical publications
Planning	Job-recording systems
	Scheduling services
	Plant-delivery systems
	Communications systems

ORGANISATION

The maintenance department should be organised in such a way that all resources, including trade skills, are available when required at the workface so that the maintenance function can be effectively and efficiently performed. Various organisational models can be adopted, depending on the size and nature of the production activity. It is strongly recommended that technicians clearly understand how their own organisation functions so that they can use this information to maximum advantage. In order to get things done it is important that the various lines of authority and responsibility are fully understood.

ROTATING MACHINERY

DEFINITIONS

A rotating machine is one in which the main working components rotate about a fixed centre in a regular manner. Most such machines incorporate additional subsidiary mechanisms such as linkages, slides, gears and reciprocating components, and many of the operating principles that apply to the rotating assembly also apply to these other elements.

Although there are many different types of rotating machines, they can all be classified into three basic groups in terms of their function.

Driving machines (engines or prime movers)

This group includes all machines whose purpose is to drive other machines. Examples include:

    electric motors
    steam turbines
    diesel engines
    petrol engines
    air motors

The common characteristic of these machines is that they convert an energy input of varying kinds into a mechanical output in the form of a rotating drive shaft.

Transmission machines

These are machines whose purpose is to transmit mechanical energy from a driving to a driven machine. Examples include:

    gearboxes
    differentials
    variable speed drives

The mechanical energy transmitted often undergoes a speed transformation and these machines often incorporate some means of drive disengagement such as a clutch.

Driven machines

These machines cannot operate independently and need to be coupled to a driving machine. Examples include:

    pumps
    compressors
    fans
    generators
    blenders
    machine tools

This group is by far the largest and includes a large number of different types of machines. The common characteristic is that the energy input is normally in the form of a rotating drive shaft while output may be in a variety of forms including kinetic or pressure energy of a fluid, electrical energy, kinetic or potential energy of solid materials, etc.

Because of the ways in which different types of machines are sometimes combined into sets, the distinction between the separate elements is not always clear. For example, some pumping units are directly coupled through a reduction gear to an electric motor with all three elements contained in a single housing. Although such an item may be treated as a single machine it does, in fact, contain all three elements of driving, transmission and driven.

From a maintenance viewpoint, it is important that the function of a machine, or the elements of a machine set, is clearly understood. Troubleshooting and fault correction depend on an ability to detect deviations from normal operation, and the assessment of operating conditions demands a knowledge of the function of a machine as well as the principles on which it operates. It is recommended that technicians make sure that they are fully familiar with the function (i.e. what the machine 'does') of the particular machinery with which they are concerned.

OPERATING PRINCIPLES

A prerequisite of good maintenance practice is a critical understanding of the principles on which the satisfactory operation of rotating machinery is based. This understanding provides the foundation of the maintenance technician's ability to diagnose and correct faults.

Although each machine is different in some way, and each engineering situation gives rise to special requirements, there are a number of conditions that are critical to the operation of all rotating machines. In Chapter 1, seven criteria were identified that can be used to determine the operating condition of a machine. They were:

Performance
Downtime
Service life
Efficiency
Safety
Environmental impact
Maintenance cost

If a machine is to perform 'satisfactorily' according to these criteria, then the following conditions must be satisfied.

Mounting

A rotating machine must be correctly mounted on a suitable foundation. If this condition is not met then operation of the machine may cause damage both to itself and to adjacent equipment. The methods by which a machine should be mounted on a foundation are covered in Chapter 3.

Mechanism

Every rotating machine must have an internal mechanism that is operable, in good repair and capable of achieving the performance required. There are many excellent books that describe the operation and maintenance of specific types of machines and the technician should consult these for information when necessary.

Balancing

The rotating components or assemblies of a rotating machine must be correctly balanced. Assemblies that are not balanced cause excessive vibration and high stresses. Not only does this condition cause rapid wear and frequent breakdown but sudden and dramatic failure may occur with dangerous consequences.

Correct balancing is essential to the safe, reliable operation of any rotating machine and is covered in Chapter 4.

Lubrication

A rotating machine, like any mechanism where relative motion of contacting parts is involved, cannot operate satisfactorily unless it is lubricated to reduce friction and wear. The principles of lubrication and the characteristics of lubricants are explained in Chapter 5.

Bearings

Every item of rotating machinery requires a set of properly maintained bearings that support the mechanism and restrain its motion with minimum resistance. Bearings are an essential component and a common element in all rotating machines. The principles of operation and maintenance of bearings are covered in Chapter 6.

Transmission

Because machines normally operate in 'sets', i.e. driving-driven or driving-transmission-driven, some means of connecting the input and output shafts of the separate elements is required. This may be accomplished in several ways and the principles of operation and maintenance of the most common of these are described in Chapter 7.

Alignment

When machines are assembled in sets as described above, it is vital that the inter-connecting shafts are properly aligned to each other. Poor alignment will cause vibration and lead to rapid wear of couplings, bearings, seals and other rotating elements. The principles of alignment and methods by which good alignment can be achieved are explained in Chapter 8.

Seals

Rotating machines usually contain a number of fluids such as process fluids, fuels, cooling water and lubricants. At the very least there will always be a lubricant present. The escape of any of these fluids must be prevented to avoid waste and the creation of hazards to personnel and the environment. Hence all machine joints and connections must be properly sealed.

As well as containing fluids within the machine, sealing also serves to prevent contamination from external sources such as dirt and moisture.

The principles of operation and maintenance of common types of seals are explained in Chapter 9.

Guards

In order to prevent injury to personnel, the minimum safety requirement on any rotating machine is that all exposed rotating elements should be guarded during operation. Guarding should be designed and installed in such a way that accidental interference with rotating elements is impossible. The construction of machine guards is usually straightforward and the maintenance technician should ensure that they are always secure and in good repair.

No matter what specific type of rotating machine is considered, these nine conditions represent the minimum requirement for satisfactory operation according to the criteria listed in Chapter 1. If proper attention is given to these conditions then there is every chance that satisfactory operation will be achieved.

MACHINERY MOUNTING

Rotating machines are frequently required to operate in 'sets' e.g. driver-driven, and therefore require a common bedplate on which the separate units can be mounted. This common bedplate provides a rigid, level structure which enables the machines to be maintained in alignment during operation. Independent machines that are sufficiently rigid and are not required to align with other equipment may be mounted directly onto a foundation.

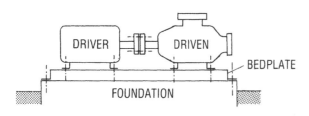

Fig. 3-1 A typical mounting arrangement for rotating machines.

It is important to realise that the mounting of the bedplate onto the foundation is critical to the operation of the machinery. If the bedplate is to provide a suitable base on which satisfactory machine operation can be achieved, then it must be level and firmly secured to the foundation without distortion. One of the functions of the foundation is to absorb machine vibrations caused by any unbalanced or inertia forces that may be present. If there is any relative movement between the bedplate and the foundation then any vibrations present, instead of being absorbed by the foundation, will tend to further loosen the bedplate mounting and cause damage to the machine. If the bedplate is not pulled down evenly and without distortion onto the foundation, then alignment of the individual machines will be difficult and casing distortion will occur when they, in turn, are pulled down onto the bedplate. This will eventually lead to rapid wear and damage to machine elements during operation.

SETTING AND LEVELLING

The setting and levelling process ensures that the mounting points for the individual machines are true and level. The bedplate should be provided with machined flats on which the machines are mounted and these should be taken as reference points in the levelling process. If level itself is not highly critical then any flat horizontal surface on the bedplate can be used. For most machinery, level can be adequately determined by using an engineers' spirit level, but for large equipment, or where special accuracy is required, then a surveyors' level may be used.

The bedplate should be mounted on shims that can be adjusted during the levelling process. Shim material should preferably be corrosion resistant and of sufficient proportions to support the weight of the machine. If the shims are crushed under the load of the machine then level and alignment will be destroyed. Shims may be either flat material or wedge shaped and should always be placed at the anchor bolt locations where the weight of the machinery is concentrated. When flat shim material is used it is recommended that it is shaped to fit around the anchor bolts.

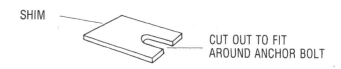

Fig. 3-2 A flat shim shaped to fit around the anchor bolt

Wedge-shaped shims provide an easy method of adjustment but care must be taken to ensure that they are installed correctly so that they are aligned with the bedplate and fully support the load. Fig. 3-3 shows the right and wrong ways of using wedge shaped shims.

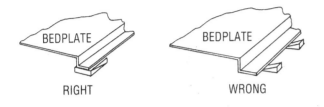

Fig. 3-3 Using wedge-shaped shims.

In order to provide some means of determining whether or not the bedplate is pulled down evenly and without distortion, it is recommended that a dial indicator is set up across the coupling, in the same way as for coupling alignment.

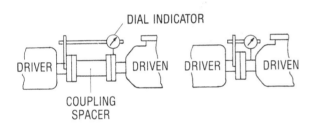

Fig. 3-4 A dial indicator set up across a coupling.

The dial indicator will allow the relative position of the machine shafts to be compared before and after the bedplate anchor bolts are pulled down. If there is a significant change in that relative position then this is an indication that the bedplate was distorted when the bolts were tightened. To overcome this problem it will be necessary to slacken the anchor bolts one at a time and use feeler gauges to determine which mounting requires extra shims under the bedplate to prevent the distortion.

Procedure for setting and levelling

1 Ensure that the foundation is clean and free of debris and that anchor bolts are correctly positioned.
2 Lower the bedplate, with the machines installed, onto the foundation and install preliminary shims under the bedplate and adjacent to each anchor bolt. The thickness of the shimpack used will vary depending on the size and weight of the machine but should range from a minimum of about 2cm (¾″) up to 5cm (2″) for large machines. Remember the shims have to carry the full weight of the loaded machine.
3 If the baseplate has machined flats to assist levelling, then mount a spirit level on the flats and check for level in both directions.

If no such levelling points are provided then select a clean, flat surface on the bedplate on which to mount the level. If levelling is critical and no machined flats are provided then it may be necessary to remove one of the machines and use the machined mounting points as a reference for levelling.
4 Adjust the shims under the anchor points until the bedplate is level in both directions. To save time and to avoid trial and error, the adjustment required can be calculated as follows:
  (i) Use feeler gauges to determine the amount of adjustment required to bring the spirit level into the true horizontal position.

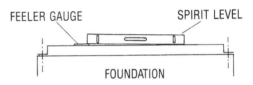

Fig. 3-5 Using a feeler gauge to determine adjustments.

  (ii) Determine the ratio of the length of the bedplate to the length of the spirit level as shown in Fig. 3-6.

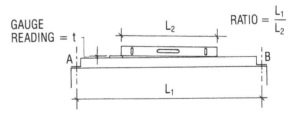

FIG. 3-6 Determining the ratio.

  (iii) Multiply the ratio by the feeler gauge measurement. This will give a figure for the adjustment required at the mounting point.

$$\text{adjustment at A} = t \times \frac{L_1}{L_2}$$

5 Tighten the anchor bolts and recheck the level. Further adjust the shims if necessary.
6 Set up a dial indicator across the machine

coupling as shown previously in Fig. 3-4. With the anchor bolts loose, set the indicator to zero at top dead centre, rotate the machine and record the indicator readings at 90° intervals.

7 Tighten the anchor bolts and again, with the indicator zeroed at top dead centre, take readings at 90° intervals by rotating the machine.

8 Compare the two sets of readings. If a deflection of more than 0.050mm (0.002″) in shaft position is measured, then some distortion of the bedplate is indicated.

9 Loosen the anchor bolts one at a time and use a feeler gauge to establish whether clearance exists between the shims and the bedplate.

10 Adjust the shims and recheck as before until the deflection recorded at the coupling is acceptable.

11 When level and deflection are satisfactory, tighten the anchor bolts making sure that all shims are correctly positioned.

The use of dial indicators is explained in Chapter 8. The arrangement used to support the dial indicator during the levelling process will also be useful when shaft alignment is carried out.

GROUTING

Once the setting and levelling procedure is complete and the anchor bolts have been tightened, the space between the bedplate and the foundation should be filled with grout, a fluid mixture of mortar-like concrete which expands when it sets. The function of grout is as follows:

- To secure the shims and ensure that they cannot be dislodged. It also protects the shims from corrosion.
- To prevent the accumulation of corrosive or hazardous liquids inside the bedplate.

Before grout can be poured, a dam must be constructed around the top of the foundation to a height of about 10–12mm (⅜″ – ½″) above the top surface of the shims.

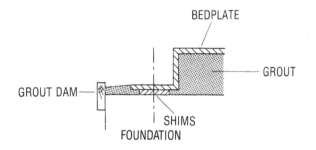

Fig. 3-7 Constructing a grout dam.

The grout is then poured inside the bedplate and thoroughly puddled to ensure that it is properly distributed and that there are no voids. Grout should be allowed to completely fill the bedplate so that there are no spaces left where liquid may accumulate.

SPECIAL MOUNTINGS

Adjustable mountings

In cases where machinery items are free-standing and may be subject to frequent changeover or change of position, anchor bolts may be replaced by adjustable mountings that provide a quick and easy means of levelling.

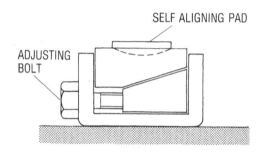

Fig. 3-8 A typical example of an adjustable mounting.

The levelling features incorporated in this type of mounting enable a machine to be levelled to precise limits in a minimum amount of time. Machines supported in this way would normally need to be equipped with flexible service lines for power, air, water, etc.

Before attempting to use this type of mounting it is important to ensure that it is clean and greased so that the necessary adjustments can be made, and so that the mounting doesn't jam. It is also important to use mountings that are designed to carry the load of the machine.

Vibration isolators

An important consideration in many machinery installations are the problems of vibration, noise and shock. To overcome these problems various types of special mountings are available. The simplest solution, if these problems are not too severe, is to insert pads of absorbent material, such as felt or rubber, between the bedplate and the foundation. In order to distribute the load the material should be sandwiched between two steel plates as shown in Fig. 3-9.

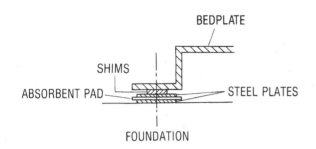

Fig. 3-9 Inserting absorbent material between bedplate and foundation

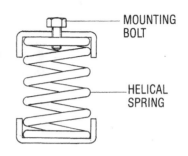

Fig. 3-10 A simple spring mounting.

Felt and other materials are available in different grades and thicknesses for different loading pressures. Where the weight of a machine is not evenly distributed over the bedplate it may be necessary to use different grades of material at different mounting points.

Where vibrations are more severe, a significant degree of deflection may be required if they are to be absorbed by the mountings. In such cases a spring type of mounting is preferred.

Various devices of this type are available ranging from a simple spring mounting of the type shown in Fig. 3-10 to a more sophisticated type which incorporates a self-damping feature.

Spring mountings normally provide a means of level adjustment which eliminates the need for shims and may also include padding to reduce noise transmission.

Vibration isolators of this type serve the dual function of preventing machine vibrations from being transmitted to other adjacent equipment and preventing the amplification of internal stresses which cause misalignment and wear of machine elements.

BALANCING

PRINCIPLES OF BALANCING

If the centre of gravity of a rotating machine element does not coincide with its centre of rotation, then the machine is said to be unbalanced. When the machine is stationary, the off-centre mass causes the machine element to settle in a fixed position. (Fig. 4-1)

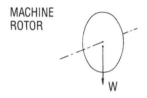

MACHINE
ROTOR

W

Fig. 4-1 Stationary machine with an off-centre mass.

As the machine element rotates, a centrifugal force associated with the off-centre mass develops and imposes a fluctuating load on the shaft support bearings as shown in Fig. 4-2.

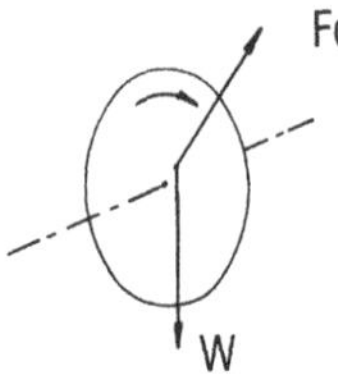

Fc = CENTRIFUGAL
FORCE

W

Fig. 4-2 Rotating machine with an off-centre mass.

The size of this force depends not only on the mass of the rotating element, but also on the extent to which the mass is off-centre and the speed of rotation. The resulting load imposed on the bearings cycles continuously through 360° with every rotation of the shaft. In addition to imposing high loads and fatigue stresses on the bearings, a fluctuating load of this type will set up vibrations that will be transmitted through the machine and the surrounding structure.

Balancing is the term used to refer to the process of improving the mass distribution of a rotating machine element, so that it rotates in its bearings without giving rise to unbalanced centrifugal forces. A machine element in which the centre of gravity and the centre of rotation coincide will settle in any position when stationary, and when rotating will not impose any additional loads on the bearings due to unbalanced centrifugal forces. The only (radial) load imposed on the bearings should be due to the weight of the rotating element.

In practice, it is impossible to achieve perfect balance and, even after sophisticated balancing techniques have been used, a rotating element will always possess some residual imbalance. The need for accurate balancing increases with the mass and size of the rotor and the speed of rotation.

To ensure safe and reliable operation of any rotating machine it is desirable that the rotating elements be balanced within defined limits according to the size and nature of the machine. It is common practice these days to monitor machine vibrations as a means of determining standards of machine balance and the technique involved is discussed in Chapter 10.

The mass distribution of a machine element can be changed by either adding or subtracting mass at any position on the rotor. Mass may be added by welding, bolting or otherwise attaching additional material to the rotor, usually in measured quantities. Mass may be removed either by drilling a hole in the rotor or by grinding or filing material from the surface. The key to successful balancing lies in the method used to determine the position at which mass is to be added or removed. In most cases it is preferable to remove material from the heavy side of a rotor, rather than add material to the light side because this overcomes the problem of having to provide a means of attaching the additional material. However, there may be circumstances where this is not possible because of strength or visual considerations.

STATIC BALANCING

The simplest and easiest method of balancing is one that uses static conditions to determine the relative position of the centre of gravity and centre of rotation. This method is relatively straightforward but it is strictly limited to machine elements with only one plane of correction, such as circular saw blades and other 'thin' rotors.

The procedure involves setting up the rotor so that it can rotate freely and settle in its equilibrium position. This usually requires the rotor to be removed from the machine and balanced on knife edges as shown in Fig. 4-3.

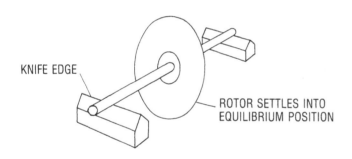

Fig. 4-3 Preparing the rotor for static balancing.

It is vital that, whatever means are used to support the rotor, there is no resistance to the rotor finding its true equilibrium position. If there is any friction present this may cause the rotor to stick and lead to mass being added or removed in the wrong position. The amount of material to be added or removed is largely determined by trial and error, although observation of the rotor as it settles to its equilibrium position will help to determine the degree of imbalance. It is necessary to check after each adjustment until satisfactory balance is achieved.

It is recommended that the rotor is first balanced by adding a lump of plasticene or similar substance at the required location. When balance is achieved this location is marked and the plasticene can be weighed. An equal amount of metal can then be added in the same position, or removed from a position 180° opposite.

DYNAMIC BALANCING

Where a rotor has more than one plane of correction, i.e., anything other than a 'thin' rotor, balancing must be carried out dynamically. It is normal practice for the machine rotor to be removed and temporarily installed on a special balancing machine. The machine is then run up to a suitable speed and the out-of-balance forces measured.

The principle of dynamic balancing is based on the measurement of the rotating couples that are set up as a result of the out-of-balance forces. Because most rotating machine elements have their mass distributed over some axial distance, the problem of determining the correct position for mass compensation becomes more complicated. Fig. 4-4 shows the twisting effect on the support bearings of the centrifugal force associated with the off-centre mass distribution.

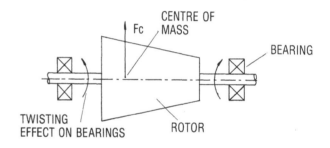

Fig. 4-4 Twisting effect associated with off-centre mass distribution in a machine rotor.

It would be a simple matter to determine the magnitude of the necessary compensatory mass by the static balancing method described above. However, the difficulty lies in determining the axial position for the mass to be located. If this position is chosen incorrectly then out-of-balance forces will continue to exist, as shown in Fig. 4-5.

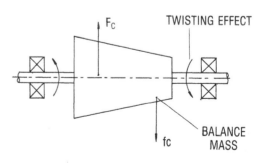

Fig. 4-5 Compensatory mass in the incorrect axial position.

Although the rotor may now be statically balanced, when it is rotated the centrifugal forces associated with the off-centre mass distribution continue to create a twisting effect on the support bearings. Because the rotor is rotating, this twisting effect fluctuates just as the original out-of-balance forces did, and proper balance is not achieved.

In a dynamic balancing machine the rotor is set up in bearings fitted with sensing devices which measure the effect of the rotating couple due to the out-of-balance mass. The information is then computed, along with other information regarding the rotor dimensions, to determine the size and correct location of the required mass compensation. As with static balancing, material is then added or removed and balance rechecked until a satisfactory standard for the particular machine is reached.

It should be recognised that, while static balancing is a relatively simple procedure that any technician can perform given the right equipment, dynamic balancing is a specialised activity normally carried out in a specially equipped workshop by trained personnel.

C H A P T E R 5

LUBRICATION

5.1 PRINCIPLES OF LUBRICATION

Whenever the surfaces of two bodies are in contact, the force of friction will resist relative motion between them. The operation of almost all industrial equipment relies on the relative motion of separate machine elements and lubrication is necessary to overcome the effects of the friction forces. To lubricate means 'to make smooth and slippery', and thus the application of a lubricant helps to reduce the effect of friction. Friction causes energy to be wasted in the form of heat and causes the rubbing surfaces to wear. The introduction of a lubricant separates the surfaces in contact and thus reduces the effects of friction although friction can never be entirely eliminated.

The basic purposes of lubrication are to:
　reduce friction
　reduce wear
　dampen shock
　cool moving elements
　prevent corrosion
　seal out dirt

FRICTION

The force of friction always opposes relative motion between two bodies regardless of the shape, size or nature of the bodies. Friction not only exists between solids but also between liquids and solids and between gases and solids. However, we are only concerned here with friction between solids.

Causes of friction

Molecular attraction
The molecules of a substance are held together by electromagnetic forces between positively and negatively charged atoms as shown in Fig. 5-1.

These forces can also act across the boundary of two substances in contact. Hence the negatively charged atoms of one substance attract the positively charged atoms of the other and vice versa. This molecular attraction increases as the surfaces in contact become smoother and theoretically two perfectly flat surfaces can not be separated once they come into contact except by mechanical means.

Fig. 5-1 Electromagnetic forces between positively and negatively charged atoms.

Interlocking of asperities
When viewed under a microscope even the smoothest surface is seen to contain a series of peaks and valleys (asperities). When two surfaces are in contact with each other, these asperities interlock and cause resistance to relative motion.

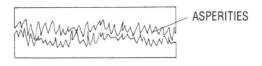

Fig. 5-2 Interlocking of asperities of two surfaces in contact.

Surface waviness
No surface is, in fact, perfectly flat but contains an element of waviness on which the irregularities

referred to above are superimposed, as shown in Fig. 5-3.

This waviness also creates resistance to relative motion in a similar way to the interlocking of the asperities.

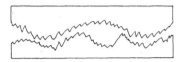

Fig. 5-3 Surface waviness.

Local welding

Because it is only the tips of the asperities that come into contact, the actual area of contact is much less than the total surface area. This means that the force between the surfaces, instead of being carried by the whole surface area, is only carried by that part in actual contact. Hence the pressure developed on these areas is extremely high and often can be sufficient to generate enough heat to cause the surfaces to melt and stick together as shown in Fig. 5-4.

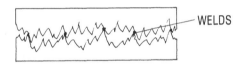

Fig. 5-4 Local welding of two surfaces in contact.

In order for the surfaces to move relative to each other, it is necessary for these welds to. be broken.

Friction force

The amount of friction developed, that is, the size of the friction force, depends on:
- The type of frictional system,
- The nature and condition of the surfaces,
- The normal force between the surfaces.

The first two factors are given by the coefficient of friction and the relationship can be expressed mathematically as

$$F = \mu N$$

where F = force of friction
μ = coefficient of friction
N = normal reaction between the surfaces

Types of friction

There are three types of friction that can exist between solids.

Static friction

When a force is applied in order to move one surface with respect to another there is a certain minimum force required before movement can be achieved. The friction force resisting motion can be calculated from the relationship $F = \mu_s N$ where μ_s is the coefficient of static friction.

Dynamic friction

Once motion occurs, the force required to maintain it reduces and is less than the force required to initiate motion. This force can also be calculated from the relationship $F = \mu_d N$ where μ_d is the coefficient of dynamic friction.

The coefficient of dynamic friction is always less then the coefficient of static friction for any two surfaces because of the increased adhesion due to local welding that occurs when the surfaces are stationary. These coefficients are determined by experiment and common values can be found from Engineering Tables.

Dynamic friction is the main concern in the case of machine elements and there are two types to consider.

Sliding friction This is the commonly understood situation giving rise to friction involving one surface sliding over another.

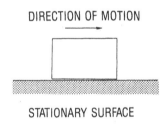

Fig. 5-5 Sliding friction.

This is the type of friction which occurs in plain bearings, cylinders, slides and cross heads, etc.

Rolling friction This is a rather different form of friction and to understand how it arises we must consider what happens when a ball, roller or wheel rolls across a surface. Although it is not usually visible to the naked eye some deformation of the opposing elements occurs with the result that a 'swell' in the surface occurs ahead of the rolling element as shown in Fig. 5-6.

The swell causes resistance to motion and this can be reduced, as with rolling element bearings, by making the surfaces as hard as possible in order to resist deformation.

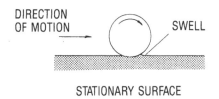

Fig. 5-6 Rolling friction.

Effects of friction

There are two major effects of friction which have consequences for machine elements and which lubrication is intended to overcome. Much of the energy that is used to overcome the force of friction is converted into heat. If this heat is not dissipated, but is allowed to build up, the moving elements may fuse together or 'seize up'.

The other effect of friction, which also uses up energy, is **wear**. The constant rubbing of one surface on another causes a physical change to occur to the topography of the surfaces and this may ultimately lead to a stage where the elements concerned can no longer perform their function effectively. The analysis of wear patterns is an important step in the troubleshooting process and in the assessment of lubricant effectiveness.

WEAR

Wear can be defined as 'the transformation of matter by use' and it results in a diminishing of the dimensions of machinery parts. Mechanical wear is the most important of four types of wear which convert useful materials into useless debris. The others are chemical (corrosion), bacteriological (decay) and electrolytic (metal removal). Mechanical wear is primarily caused by the relative motion of surfaces in physical contact and is a complex process often occurring as a result of some combination of the following:

Adhesive wear

Two surfaces in contact bear on the tips of the asperities which results in local welding and causes surface wear. When relative motion takes place a shearing process occurs. Sometimes the local welding is so strong that instead of shearing taking place at the surface, grain displacement occurs, creating pits or craters. This is known as **galling**. Adhesion is the most fundamental type of wear and occurs according to the nature of the opposing surfaces.

Abrasive wear

Abrasion occurs when particles of hard material become trapped between the surfaces or when hard particles are embedded in or attached to an opposing surface. The hard particles remove metal from the softer material rather like a grinding operation. The hard particles may be present in the materials, or may accumulate due to the shearing action between the surfaces. Abrasive particles may also enter the space between the surfaces from external sources.

Fatigue

Fatigue failure is caused by repetitive or cyclic stresses which weaken a material and cause failure to occur at a level of stress well below the normal strength of the material. This can happen in rolling element bearings due to the cycling effect of the rolling elements, and in plain bearings due to fluctuating load patterns. Fatigue first affects the sub-surface region and ultimately causes cracks to appear on the surface of the material. As the process progresses the surface begins to flake off. When this condition becomes severe it is known as **spalling**.

Fretting

Fretting, often known as false brinelling, is a type of corrosion which occurs when there is a slight, imperceptable motion between two surfaces in contact. The source of such motion is usually vibration. The surfaces at the area of contact deteriorate and break down to form oxides. This problem is sometimes encountered when machinery is in transit and the vibration of the carrier is transmitted to rolling element bearings. The slight motion between the rolling elements and the raceway can be sufficient to cause fretting to occur. The problem of fretting can be solved by eliminating the relative motion or plating the surface with non-oxidising materials, rather than by lubrication. In the example given above, barring the machine over from time to time may avoid the problem.

THEORY OF LUBRICATION

The purpose of lubrication is to reduce the friction between two contacting surfaces in relative motion by introducing a lubricant between the surfaces. In theory, any substance, whether it be liquid, solid or gaseous, may act as a lubricant, although in practice only a limited number of materials have the properties necessary to act as effective lubricants (see Section 5-2).

Lubricants work by creating special conditions between the surfaces in contact. There are four main types of lubricant film conditions that can exist:

Dry friction

When surfaces are clean and dry and no lubricant exists, the condition is referred to as dry friction. This condition gives rise to the greatest frictional resistance to motion.

Boundary lubrication

In this condition a thin layer of lubricant is present but significant metal-to-metal contact still exists. Hence part of the load is taken by the lubricant but most is still taken by the surface high spots. This condition, when used to combat heavy loading, is known as extreme pressure lubrication.

Mixed-film lubrication

This is an intermediate condition between boundary and full-film lubrication in which the lubricant layer is thicker than in boundary lubrication but some metal-to-metal contact still exists.

Full-film lubrication (also called thick-film)

This is the condition in which the moving surfaces are completely separated by the lubricant film. This can occur in three different ways:

Hydrostatic When the lubricant is supplied under pressure from an outside source e.g. a pump or gravity feed.

Hydrodynamic In which the pressure develops due to the resistance of the lubricant itself. This is the type of lubrication that occurs in a plain bearing and can best be understood by examining how the lubricant film forms in that situation. Fig. 5-7 shows the stages involved.

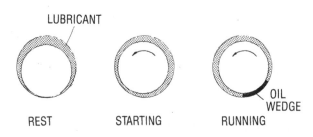

Fig. 5-7 Stages in the formation of lubricant film in a plain bearing.

While at rest the journal sits on the bottom of the bearing as shown. On start-up it climbs up the side of the bearing and establishes a lubricant film which, as it develops, forces the journal over to the other side of the bearing where it rides on a wedge of lubricant. As the lubricant is drawn into the wedge by the action of the journal it is compressed, and the pressure developed keeps the metal surfaces apart.

As the velocity of the journal increases, the wedge builds up and forces the journal and bearing closer to a concentric relationship.

Elastohydrodynamic This form of lubrication is associated with rolling friction and involves a combination of the effects of the elasticity of the surfaces in contact and the lubricant characteristics. When a ball or a roller contacts a raceway, both surfaces deform slightly but return to their original shape when the element moves on, due to the elasticity of the materials. As the rolling element climbs out of the depression lubricant is drawn into the space between the element and the raceway, and a hydrodynamic film is formed similar to that in a plain bearing.

In engineering applications such as plain and rolling element bearings it is desirable to achieve full-film lubrication to provide maximum separation of the surfaces. Where loads are very high this may not be possible and the use of extreme pressure lubricants may be needed in order to maintain boundary lubrication.

For components where positional accuracy is essential, such as ways and slides, full-film lubrication is undesirable and boundary lubriction is preferred because the metal-to-metal contact involved ensures more precise positioning of the components.

LUBRICANT SELECTION

The selection of a lubricant is determined by the following factors:

Load The load on the bearing will determine the pressure that the lubricant will have to work against.

Speed As operating speeds increase the lubricated surfaces will tend to wear faster.

Temperature The operating temperature may affect the properties of the lubricant.

Environment The lubricant may be required to cope with the presence of water or corrosive materials.

Lubricant selection should normally be left to those who are expert in the field. However, as a general rule it is worth remembering that for plain journal bearings:
- For light loads and high speeds – use a lubricant of low viscosity; and
- For high loads and low speeds – use a lubricant of high viscosity.

The decision of whether to use oil or grease as the lubricant will depend on the operating conditions.

The following comparative advantages should be taken into account:

Oil: provides cooling
feeds more easily and can be fed from a central supply
washes away dirt
can also lubricate other elements such as gears
absorbs less torque

Grease: allows simpler bearing designs
provides better sealing against dirt
is easier to contain and seal
allows longer periods without attention

METHODS OF APPLICATION

The golden rule of lubrication is said to be: 'Good lubrication depends on the right lubricant being available in the right quantity at the right time.' For this to be achieved the technician must be aware of a number of basic principles governing the application of lubricants.

- Cleanliness is vital. Lubricating equipment must be kept free of dirt and other contaminants.
- Lubricants are not necessarily interchangeable and as a general rule should not be mixed. Before changing lubricant the equipment should be cleaned out.
- An excess of lubricant, especially grease, will cause excessive heat to build up and eventual breakdown of the lubricant.
- Lubricant filters or strainers should always be changed at the recommended time.
- The selection of lubricant for a particular application should be left to qualified personnel if possible.
- Inadequate lubrication can often be identified by the operating condition of a bearing, especially its temperature. As a general rule, if a bearing is too hot to hold a hand on it, then lubrication may be inadequate and should be investigated.
- Lubricants are potentially hazardous materials and should be stored with regard to safety and effect on the environment.

There are four basic methods by which lubricants can be applied and these are selected according to design criteria and the particular demands of the equipment.

Manual application

Whether the lubricant is liquid, semi-solid or solid, the simplest method of application is by hand. An oil can may be used for liquid lubricant, a grease gun for grease and a brush or spray gun for solid lubricant.

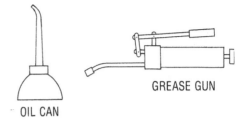

Fig. 5-8 Equipment for manual application of lubricant.

Gravity

This method is only suitable for liquid lubricants and is sometimes referred to as drip-feed oiling. There are various types of drip-feed oilers and they usually include some method of feed regulation.

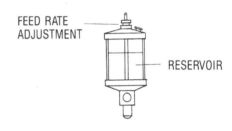

Fig. 5-9 Drip-feed oiler.

Splash lubrication

Splash lubrication relies on the components requiring lubrication being partially immersed in an oil sump so that they pick up oil as they rotate. The oil picked up in the process may also be deposited on the shaft bearings and other components. A variation on this method is the ring-type oiler which uses a steel or brass ring which rotates with the shaft and picks up oil which it deposits on the upper surface of the shaft. Examples of these methods are shown in Fig. 5-10.

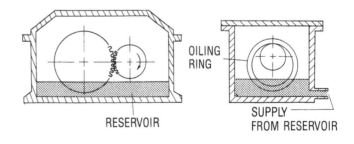

Fig. 5-10 Splash lubrication methods.

Pressure lubrication

Many industrial applications, especially where loads are heavy and operating speeds are high, require a pressurised system to ensure that an adequate supply of lubricant can be maintained. This usually takes the form of a circulating system such as that shown in Fig. 5-11.

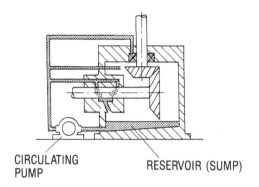

CIRCULATING PUMP   RESERVOIR (SUMP)

Fig. 5-11 Circulating system of lubrication.

5.2 LUBRICANTS

CHARACTERISTICS AND PROPERTIES

There are four basic types of lubricant:
  liquid
  semi-solid or plastic
  solid
  gaseous
and they can be classified according to their source:
  animal
  vegetable
  mineral

Most lubricants are mineral based and are obtained from petroleum by refining processes and further purification and blending. Petroleum and petroleum products belong to the group of chemicals known as 'hydrocarbons' because they are compounds of the elements hydrogen and carbon in varying combinations.

Animal fats come from common animals such as cattle, sheep, pigs and fish and are melted down to remove impurities. Vegetable oils are produced by squeezing vegetables and seeds (e.g. soyabean and rapeseed) to produce pulp and juice and then further refined to remove impurities. Animal and vegetable oils are less stable than mineral oils and break down more easily.

Lubricants are classified according to their properties and a number of standard tests are applied to common types of lubricants, such as oils and greases, in order to determine their properties.

Criteria for measuring the properties of oils

The properties of lubricating oils are measured by the following criteria:

Viscosity This is the single most important characteristic and refers to the 'thickness' of a fluid and is also described as resistance to flow. Viscosity is affected by temperature and decreases with increasing temperature. There are various ways of measuring viscosity, all of which are based on the time taken for a fixed volume of oil to pass through a standard orifice under standard conditions. The SI unit is the centistoke (cSt).

Viscosity index The rate of change of viscosity with temperature is known as the viscosity index. It is normally desirable for the viscosity of a lubricant to remain the same over a wide range of temperatures. A high viscosity index indicates such a property whereas a low index indicates that the oil tends to thin out rapidly with increasing temperature.

Flash point This is the temperature at which the vapour of a lubricant will ignite.

Fire point This is the temperature, higher than the flash point, required to form sufficient vapour from the lubricant to cause it to burn steadily.

Pour point This is the low temperature at which the lubricant becomes so thick that it ceases to flow.

Oxidation resistance When hydrocarbons are exposed to the atmosphere, especially at increased temperatures, they tend to absorb oxygen. This causes a chemical change in the oil that makes it useless for lubricating purposes.

Emulsification An emulsion, e.g. a mixture of water and oil, is undesirable because it has poor lubricating properties. Emulsification is a measure of the tendency of an oil to mix intimately with water. Demulsibility measures the readiness of an oil to separate from water in an emulsion.

Criteria for measuring the properties of greases

The properties of greases, being semi-solid rather than liquid, are measured by a separate set of criteria:

Hardness Because greases are semi-solid they can be considered as ranging from hard to soft. These ratings are based on the results of a penetration test and the standard gradings used by the National Lubricating Grease Institute (US) are as follows:

NLGI No.	Consistency	ASTM worked penetration at 25°C (77°F) 10^{-1}mm
000	very fluid	445–475
00	fluid	400–430
0	semi fluid	355–385
1	very soft	310–340
2	soft	265–295
3	semi firm	220–250
4	firm	175–205
5	very firm	130–160
6	hard	85–115

Dropping point This is the temperature at which the grease will change from semi-solid to liquid i.e. the melting point.

Pumpability This is a measure of the ease with which the grease will flow through a system.

Water resistance This determines whether or not a grease will dissolve in water. This property is important where there is a likelihood of water coming into contact with the lubricant.

Stability This property determines the ability of a grease to retain its characteristics with time. Some greases become soft and thin after being in use for a while.

LIQUID LUBRICANTS

Rather than classifying lubricants according to their composition, it is more useful for industrial purposes to classify them according to their application. There are several commonly recognised categories.

Circulating oils

These oils are designed to circulate through a closed system such as a crankcase or hydraulic circuit. Because they remain in the system for some time they normally include a range of additives such as anti-oxidants, anti-foam agents, rust inhibitors, etc.

Gear oils

Gear oils are required to handle the relatively high pressures that develop between gear teeth and also to dampen the shock of impact. They are generally required to have high viscosity although they are often required to lubricate bearings as well and must be able to transfer heat. They are normally classified as follows:

Grade	Viscosity (cSt)
Light	140–160
Medium	200–240
Heavy	420–500
Light E.P.	60– 75
Heavy E.P.	300–360
Cling type gear shield	200–240

The oils in the first group are designed for gears running in enclosed spaces and lubricated by splash or pressure systems. The extreme pressure type of oils are particularly used where tooth loads are high, such as hypoid gears used in automotive transmissions. Cling type oils are used for open gears and are specially compounded to resist being thrown off as the gears move. Gear oils often contain anti-wear additives where tooth loads are expected to be heavy.

Machine and engine oils

These are general purpose oils suitable for once-through systems and were originally designed for the external operating parts of machinery that could be oiled with cans or cups. They are commonly used on plain bearings and for slides and ways. Viscosities range from 35–200 cSt. They should not be used in situations where sludge may form.

Spindle oils

These are very carefully refined, high quality oils designed to lubricate the spindles of the textile industry and also to lubricate delicate instruments and other sensitive equipment. They are usually straight mineral oils with low viscosities ranging from 1–25 cSt.

Refrigeration oils

Oils used for lubricating refrigeration equipment are special straight mineral oils with low pour point and free of wax and moisture. They range in viscosity from 15–120 cSt and are usually non-foaming.

Steam cylinder oils

These are special purpose oils, formulated by compounding mineral oils with animal oils, and sprayed into the steam or applied directly to the walls of steam cylinders. The animal oils help to ensure that they adhere to the surfaces they are supposed to lubricate. Viscosities are high and range from 20–35 cSt at 100°C (210°F). [500–1300 cSt at 38°C (100°F)].

Special purpose oils

A number of special purpose oils are available for particular applications, not necessarily associated with lubrication. These include:
   fire resistant oils
   cutting oils
   heat treatment oils

General purpose oils

In recent years manufacturers have developed a range of multipurpose oils which meet multiviscosity or multigrade specifications. These are primarily used for lubricating automotive engines and transmissions and can withstand the variation between summer and winter conditions.

The above categories are only general and many oils are suited to more than one particular application. The categories are listed merely as a guide and the selection of lubricants should be referred to the supplier.

GREASES

Categories of greases

Soap-thickened mineral oils

These are the most commonly used greases and are classified according to the type of soap base used to thicken the oil.

Calcium-based The most important general purpose greases. They do not dissolve in water and are limited to applications below 70°C (160°F).

Sodium-based General purpose greases with a high dropping point of 150°C (300°F). They can be used for parts near a heat source.

Barium-based General purpose greases suitable for temperatures up to 135°C (275°F) but less suitable for low temperature applications due to the high soap content.

Lithium-based Have good water resistance and stability and are suitable for both low temperature −50°C (−60°F) and high temperature 150°C (300°F) applications.

Strontium-based Have good resistance to water washing and to corrosion from substances such as salt water. They can also operate up to 175°C (350°F).

Aluminium-based Have the particular advantage that they adhere well to the surfaces they are supposed to lubricate.

Complex Greases whose basic ingredients have been fortified, modified or treated in some way in order to achieve high performance in a particular application.

Mixed-base Combine the advantages of two or more different soap bases.

Multipurpose The advantage of these greases is that they reduce the number of types of grease required in a particular industrial situation and also reduce the possibility of using an inappropriate grease. They are usually either barium or lithium-based and combine the properties of specialised greases.

Mineral oils mixed with solids

These are heavy greases used to lubricate rough-fitting components which operate under high pressure or at low speed such as conveyor systems.

Heavy asphaltic-type oils

Although these are really thick oils they are classified as greases and are used to lubricate open gearing and wire ropes. They are painted on when hot, and cool to form a protective coating.

Extreme pressure greases

These are designed to give improved load carrying capacity in rolling element bearings and gears. Various types of additives are used to improve the film-strength (resistance to rupture) of the grease. They are often lithium-based and also contain further additives to improve lubricity (oiliness).

Roll-neck greases

These are specialised greases used for lubricating plain bearings in rolling mill equipment and are usually used in block form which is cut to shape to fit the bearing.

Synthetic oil-based greases

Various synthetic fluids are used to produce greases with special characteristics not easily attainable with mineral oils. The fluids used include: chlorofluorocarbon polymers, dibasic acid esters, polyglycols, silicones and polyphenyl ethers.

The silicone-based greases are particularly useful because they are suitable for a wide temperature range and are resistant to oxidation. For this reason they are often used in sealed-for-life bearings.

Applications of greases

Ways and slides

The greases used for ways and slides are usually sodium-based of consistency NLGI No. 1 or 2.

Plain bearing greases

Greases can be used for the lubrication of plain bearings operating at low speeds up to about 400 RPM. Grease has a tendency to work its way out of the ends of the bearing so it must be regularly replaced, especially as speeds increase. Most of the soap-based greases can be used for lubricating plain bearings. Lithium-based grease is one of the most versatile.

Rolling element bearings

Greases used for rolling element bearings must be free of contaminants and chemically active elements that may attack the bearing or the seal, and they must also maintain consistency in operation. Most general purpose greases are suitable depending on the particular conditions.

SOLID LUBRICANTS

Solid lubricants, such as graphite and molybdenum disulphide, may be used as additives for oil and grease or may be used alone in their dry state where oils and greases do not perform satisfactorily. Solid lubricants may have advantages over oils and greases where:
- Bearings are inaccessible or likely to be missed by routine maintenance;
- The use of liquid lubricants may cause product contamination;
- There is a tendency for galling or seizing to occur;
- Operating temperatures are either too low or too high for oils and greases;
- Fretting corrosion is a possibility;
- Vacuum conditions exist which tend to cause evaporation of most substances;
- The lubricant is exposed to nuclear radiation

which causes ionisation of organic compounds and change in viscosity.

The most satisfactory method of application for solid lubricants in the dry state is spraying or dipping. They may also be brushed on but care must be taken to ensure that film thickness is even. When applied in this way they are usually used in a bonded form using a 'binder' which helps the lubricant to adhere to the surface. Phenolic-resins are often used for this purpose.

Molybdenum disulphide in its powdered form is often used on metal forming dies and on threaded parts. Bolts that have been lubricated can be tightened closer to strength limits and undone more easily.

ADDITIVES

The wide variation in the requirements of modern machinery is beyond the ability of straight oils and greases to handle and a range of additives has been developed to improve lubricant properties. The most common types of additives are:

Oxidation inhibitors These are designed to prevent the chemical breakdown of lubricants and the formation of acids.

Detergents and dispersants Detergents help to keep surfaces clean by preventing the formation of dirt particles. Dispersants keep the dirt particles in suspension by enveloping the dirt particles and preventing them from adhering to the metal surfaces.

Rust and corrosion inhibitors These prevent or retard the formation of rust and also protect metal parts against corrosion by contaminants.

Pour point depressants These are used to ensure that the lubricant will maintain its ability to flow at low temperatures.

Viscosity index improvers These are added to reduce the effect of changes in temperature on viscosity.

Anti-foam agents These help to break up the air bubbles that tend to form in a circulating or a hydraulic system.

Anti-friction compounds Oiliness and lubricity of the lubricant are increased in order to reduce the coefficient of friction between the rubbing surfaces.

Anti-wear agents These also reduce friction and wear due to scoring, seizing and rubbing.

Extreme pressure agents These are used primarily in gear oils and help to cushion the shock between gear teeth at high loads.

Emulsifiers These keep water away from contact surfaces by forming an oil film around water particles.

Emulsion breakers When an emulsion is undesirable an emulsion breaker is used to help the oil and water separate more easily.

Adhesive compounds These ensure that lubricants adhere to surfaces and prevent them being thrown off by centrifugal force or turbulence.

A range of organic and inorganic substances is used to achieve these various effects and individual manufacturers of lubricants have their own particular formulae.

BEARINGS

A bearing is a device which supports a rotating shaft or spindle or guides one component which slides over another. In addition to its supporting function, a bearing is designed to allow relative movement between two separate components to occur with the least possible frictional resistance. Almost all industrial mechanisms contain elements which require relative movement between contacting surfaces and therefore include some sort of bearing.

In principle, bearings fit into two main categories: **Plain bearings** in which the surface of one component slides over the surface of another and where the surfaces in contact are specially prepared in order to minimise friction and wear.

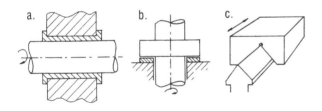

Fig. 6-1 Examples of plain bearings.

Rolling element bearings in which a series of rolling elements, i.e. either balls or rollers of various shapes, are interposed between the two surfaces in order to facilitate movement of one with respect to the other. Rolling element bearings are sometimes referred to as anti-friction bearings because the relatively small contact area of the rolling elements helps to reduce, though not eliminate, the resistance to relative motion.

Bearings can also be classified according to the type of function they perform:
Journal bearings which support a rotating shaft or spindle and confine radial motion as in Figs. 6-1a and 6-2a.

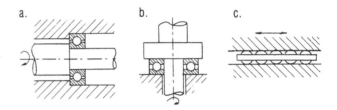

Fig. 6-2 Examples of rolling element bearings.

Thrust bearings which prevent axial motion of a shaft as in Figs. 6-1b and 6-2b.
Linear bearings which guide or support relative motion between components in a straight line as in Figs. 6-1c and 6-2c.

The function a bearing performs is largely determined by the type of load that it has to carry. The type of load also determines the particular type of bearing selected. Loads can be one of three kinds:
Radial
Axial
Combination

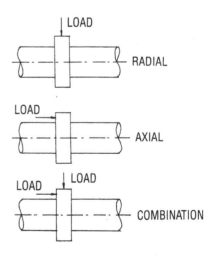

Fig. 6-3 Bearing loads.

6.1 PLAIN BEARINGS

PRINCIPLES OF OPERATION

The operation of a plain bearing relies on relative motion between plain surfaces which are either cylindrical, in the case of journal bearings, or flat, in the case of thrust or slider bearings. In order that motion may occur with minimum resistance the effect of friction between the two surfaces must be reduced as much as possible. This is achieved by ensuring that the two surfaces are smooth and by introducing a film of lubricant between them. In some cases a bearing material which has self-lubricating properties may be used, otherwise a separate supply of oil or grease must be provided. For the best operation it is preferable that full-film lubrication is achieved so that metal-to-metal contact is eliminated. In some cases however, this may not be possible. When loads are very high it is impossible for a lubricant to maintain separation of the surfaces and even with extreme pressure lubricants, only boundary lubrication can be achieved. In the case of slider bearings, where positional accuracy of the sliding element is required such as for machine tool carriages, full-film lubrication is undesirable because the lubricant film tends to vary in thickness. Hence boundary lubrication, which allows some metal-to-metal contact and thus more accurately locates the sliding element, is preferable.

The material of a plain bearing is always softer than the material of the shaft or slider it supports. This allows the bearing to wear out rather than the shaft or slider. The design of plain bearing assemblies is usually such that the bearing material itself can be relatively easily replaced.

The key to satisfactory operation of a plain bearing is the maintenance of the correct lubricant film. This is often assisted by machined grooves in the bearing surface which act as reservoirs and ensure even distribution of lubricant across the bearing surface. The grooves are usually connected to the lubricant inlet which is often located just ahead of the most highly loaded area of the bearing.

One of the most important aspects of bearing operation is the control of temperature. This may be achieved either by ensuring an adequate flow of lubricant which acts as a coolant or by surrounding the bearing with chambers in the housing through which cooling water can circulate.

TYPES AND MATERIALS

Journal bearings

A plain journal bearing is essentially a tube which encircles, either totally or partially, a rotating shaft.

The tube is normally held in a fixed housing and provides a support for the shaft and allows it to rotate with minimum resistance.

There are two principal types of plain journal bearings – solid and split.

A solid bearing is usually called a bushing when it is of the thin-walled type i.e. up to 2mm (⁵⁄₆₄″) for a ⌀25mm (1″) bearing, and a sleeve bearing when it is of the thick-walled type.

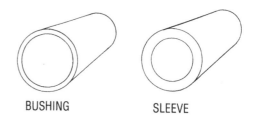

BUSHING   SLEEVE

Fig. 6-4 Types of solid journal bearings.

There are various types of split bearing but the most common consists of two halves which together form a full circle. They are used particularly for ease of shaft removal which can be accomplished by removing the top half of the bearing only.

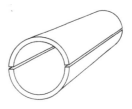

Fig. 6-5 Split journal bearing.

Similar to a split half bearing is a multipart bearing which is composed of more than two segments which fit together around the journal to form a full circle.

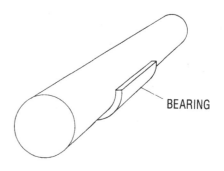

BEARING

Fig. 6-6 Partial bearing.

A partial bearing is one which supports the shaft from the bottom only and may comprise a half-circle or less. This type of bearing can only be used when loads are relatively heavy and are constant in direction.

Thrust bearings

A plain thrust bearing supports the end thrust of a shaft and restrains axial movement. The bearing surface itself consists of a stationary surface in the form of a pad against which the rotating element bears.

The most common form of plain thrust bearing is the thrust collar arrangement with the bearing face machined with radial grooves as shown in Fig. 6-7.

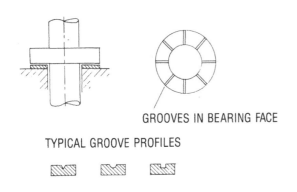

GROOVES IN BEARING FACE

TYPICAL GROOVE PROFILES

Fig. 6-7 Plain thrust bearing.

Lubricant must be fed into the bearing at the inner diameter so that lubricant flows outwards across the bearing face. Running speeds are limited by lubricant viscosity.

A slight modification to the above is a profiled pad thrust bearing where the pads are shaped to provide a convergent lubricant film and help to form a hydrodynamic wedge. The pad may be profiled to be uni-directional or bi-directional as shown below in Fig. 6-8.

UNI-DIRECTIONAL   BI-DIRECTIONAL

Fig. 6-8 Profiled pad thrust bearing.

Tilting pad thrust bearings are able to accomodate a large range of speeds, loads, and lubricant viscosities. This is achieved by providing pivoted pads which can assume a small angle with respect to the rotating collar surface, and this makes sure that a full hydrodynamic film is maintained between the surfaces.

Fig. 6-9 Tilting pad thrust bearing.

Linear bearings

Linear bearings, which are often called ways or guides, support either linear or curvilinear motion and are often associated with reciprocating mechanisms.

There are various forms that linear bearings can take, some of which are shown in Fig. 6-10.

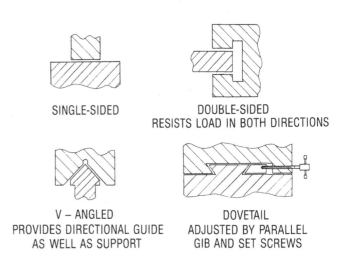

SINGLE-SIDED   DOUBLE-SIDED
                     RESISTS LOAD IN BOTH DIRECTIONS

V – ANGLED   DOVETAIL
PROVIDES DIRECTIONAL GUIDE   ADJUSTED BY PARALLEL
AS WELL AS SUPPORT   GIB AND SET SCREWS

Fig. 6-10 Typical plain linear bearings.

Materials

The requirements of a plain bearing material are complex and, in some senses, conflicting. For example, a bearing material must be hard enough to resist wear and yet soft enough to conform to the contours of the journal during the running-in process. In order to meet these complex requirements, a range of special materials that are used purely as bearing materials has been developed.

A plain bearing material must exhibit the following important characteristics:

Compressive strength To support uni-directional loading without deformation.

Fatigue strength To resist fluctuating dynamic loadings.

Embeddability To embed foreign matter and protect the journal from wear and scoring.

Conformability To tolerate misalignment and deflection under load.

Compatibility To tolerate momentary metal-to-metal contact without seizure.

Corrosion resistance To resist attack due to water or oxidised lubricant products.

The following materials are commonly used:

Whitemetals

The whitemetals or babbitt alloys are the best known of all bearing materials. They are either tin-based or lead-based alloys and contain a significant percentage of antimony. They are normally used as a coating or liner backed by steel, cast iron or bronze. Whitemetals have excellent characteristics in all respects except that compressive and fatigue strength falls rapidly with rising temperature.

Copper-lead alloys

These provide strength and fatigue resistance up to four times that of whitemetals. Conformability and embeddability are lower however, and are sometimes improved by overlaying a thin layer of whitemetal in a tri-metal construction as shown in Fig. 6-11.

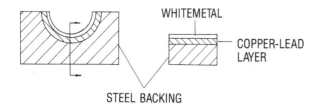

Fig. 6-11 Overlaying a copper-lead alloy with whitemetal.

Bronzes

Bronzes can be produced with a wide range of properties and are probably the most economical of all bearing materials. Four types are used.

Lead bronze Can be used without a steel backing but has low comformability and must be accurately aligned. It is easily cast and machined and is used for moderate speeds and loads.

Tin bronze Has relatively high hardness and is used in applications where loads are high but speeds are low. Tin bronzes require reliable lubrication and a hardened journal.

Aluminium bronze Has good shock and wear resistance and can be used at temperatures of 260°C (500°F) and above. It is best used in heavy duty, low speed applications.

Phospor bronze Also useful in heavy load, high temperature applications.

Aluminium alloys

These may be used as solid aluminium on steel or with an overlay of whitemetal. They have high load

capacity, good fatigue strength and excellent corrosion resistance, but poor conformability and embeddability. They are best suited to heavy loads at moderate speeds.

For more specific details regarding the composition of bearing materials it is necessary to consult either the appropriate standards or manufacturer's information.

MAINTENANCE PRACTICES

General

There are a number of general considerations that relate to the maintenance of plain bearings.

- Cleanliness is a keynote that must be observed when handling bearings of any kind. Dust and dirt particles should be kept away from bearing surfaces and all bearing components should be kept dry and protected at all times.
- Good alignment is essential to plain bearings without self-aligning properties.
- When plain bearings are disassembled all parts should be marked so that they can be reinstalled in their original positions. The matching of parts during running-in makes them unsuitable for operation in other positions.
- When fitting plain journal bearings careful attention should be paid to the correct tensioning of bolts. Bolting operations are normally conducted by manufacturers with bolts tensioned to the same specification recommended in their maintenance instructions. Incorrect tensioning may affect clearances and bearing crush.
- The key to good bearing performance is having the correct clearances and the correct lubricant.
- Bearing surfaces should be given preliminary lubrication before the shaft is turned.
- The journal surface or thrust collar should always be checked to ensure that it is perfectly smooth. In the case of a journal, a micrometer should be used to check the size so that replacement bearings will have the correct i.d. (inside diameter).

Assembly

Bushes and sleeves

Bushes and sleeves may be either a press-fit or a shrink-fit in the housing.

In the case of a press-fit the following procedure should be adopted.

1 Check the dimensions of the bush and housing and compare with manufacturer's recommendations to ensure they are correct.
2 Ensure that the mating surfaces are clean, smooth and free of burrs.
3 Press the bearing into the housing using an arbor

press if possible. Hammering should be avoided as this may ruin the bearing. The bearing should be pressed in in one continuous motion and a lubricant should be used.

4 Check the inside diameter of the bush to see if it has been reduced due to 'close-in' as a result of the press fit.

5 Hone or ream the bush to the recommended size.

In the case of a shrink-fit, the bearing can be assembled either by heating the housing or by cooling the bearing. If the housing is heated, care should be taken to ensure that heating is uniform so that no warping or distortion occurs. Heat up should not be too rapid and normally a temperature of around 100°C should provide sufficient expansion. Avoid heating the housing any more than is absolutely essential. If it is not convenient to heat the housing, the bearing may be packed in dry-ice or chilled in a deep-freeze unit before assembly. As in the case of a press-fit, it is advisable to check the inside diameter of the bearing when temperatures have returned to normal to ensure that it has not been reduced due to 'close-in'.

Split bearings

There are a number of important considerations to be taken into account when assembling split bearings of the precision type.

- It is vital that the surfaces of the bearing and housing are clean. Especially check between the bearing and housing and the joint faces of both the bearing and housing. Also check the oil-ways and oil grooves.

- A split bearing is provided with 'free spread' to ensure that both halves assemble correctly and do not foul the shaft along the joint line when bolted up.

Fig. 6-12 Free spread in a split bearing.

Free spread can be adjusted by tapping the bearing gently with a wooden mallet as shown in Fig. 6-13.

Manufacturer's recommendations should be checked to determine the amount of free spread allowed.

- To ensure that the bearing inserts have good

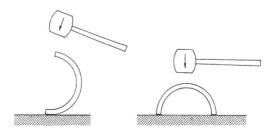

Fig. 6-13 Adjusting free spread.

contact with the housing they are made so that the inside diameter at right angles to the parting line is slightly larger than the diameter across the parting line as shown in Fig. 6-14.

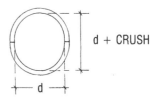

Fig. 6-14 Bearing crush.

This excess is called crush and ensures that when the bearing is bolted up it will be held securely and in full contact with the housing.

The bearing halves should be installed in the housing so that the crush is even on both sides of the housing. Bolts should be torqued up progressively and evenly on both sides.

Clearances

The correct adjustment of clearances is vital to the operation of plain bearings. If clearances are insufficient then metal-to-metal contact will occur causing excessive wear. If clearances are too great then the hydrodynamic wedge will break down and also cause metal-to-metal contact.

In the case of journal bearings there are several ways in which clearances can be checked.

- Measurement of the inside diameter of the bearing and outside diameter of the journal can be made with either a micrometer or gauges. Care should be taken if using a micrometer, to take readings in several positions.

- A soft material that will deform easily, such as plastic thread, can be inserted between the bearing and the journal before the bolts are torqued up. After the bolts have been torqued and released again, the flattened material can be compared with a suitably prepared chart and the clearance read directly.

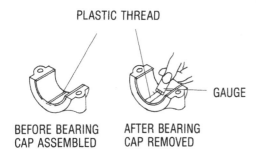

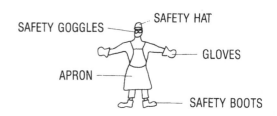

Fig. 6-15 Checking bearing clearance with plastic thread.

Fig. 6-16 Protective clothing and safety glasses.

- A third alternative is to insert lead or brass shims in between the bearing and the housing. The shims are adjusted to the point where the shaft turns easily but where an extra 0.025mm (0.001″) added to the thickness of the shim would cause it to jam. Great care has to be taken not to damage the bearing surfaces if this method is used.

If, whatever method is used, some out-of-roundness is found to exist, then it is the minimum clearance which is critical.

In the case of thrust bearings it is also important to measure the end float to ensure that sufficient clearance exists between the thrust collar and the bearing. This can be done by forcing the shaft hard up in each direction and either measuring the clearance at the thrust faces with gauges or shims or setting up a dial gauge and reading against a suitable shoulder on the shaft.

Relining journal bearings

Although it is less common these days for plain bearing liners to be poured on site it is often still practical for large bearings used in heavy equipment. For smaller bearings, spare liners are usually purchased from the manufacturer and held as stock. However the technique of relining is still used and should be understood.

As a general rule only bearings with a low melting point, such as whitemetal, are relined. Copper and aluminium based alloys have high melting points and the higher pouring temperatures tend to cause distortion of the bearing supports.

The following procedure for relining is recommended:

1 Before starting the operation ensure that proper protective clothing and safety glasses are available. The process is one of casting a molten metal and this may spit or splash during pouring.
2 First clean up the old shell by heating with a torch to remove all traces of the old whitemetal liner. Care should be taken not to overheat the liner and cause distortion.

3 In order to be properly tinned the steel shell should be chemically clean and slightly etched. Clean first in a metal cleaning solution, rinse and then dip into an acid solution for a few minutes. Give a final rinse with clean water.
4 Once cleaned, immerse the shell in hot soldering flux at 65°C (150°F).
5 Next tin the shell by dipping it into molten tin at 285°C (550°F) for steel and cast iron or 50% tin 50% lead solder for bronze liners.

Hold the shell submerged just long enough for it to reach the bath temperature. Examine the tinned surface and if any areas have not tinned completely, wire brush them and reflux and reimmerse them.

N.B. Cast iron is difficult to tin because of the presence of graphite and absorbed oil. This can be overcome by burning off and brushing, or by preparation in a molten salt bath.
6 If the bearing is too large or a tin bath is not available then tinning may be accomplished by heating with a blowtorch or gas flame and using a stick of tin or solder.
7 Set up the bearing shell in a fixture comprising a plate and mandrel or core as shown in Fig. 6-17.

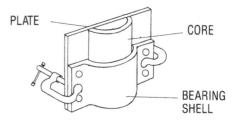

Fig. 6-17 Setting up the mould.

If the bearing is too large to be removed and must be lined on site then register plates must be set up at each end to form the inside radius as shown in Fig. 6-18.

In some cases it may be convenient to use the journal as a mandrel in which case it should be coated with graphite to prevent adhesion of the whitemetal.

BEARING SHELL

REGISTER PLATES

Fig. 6-18 Using register plates to form the inside radius.

The bearing shell should be preheated to 120°C (250°F) immediately before the liner is poured.

8 The whitemetal should be heated to the recommended minimum pouring temperature as given by the relevant standards or by the manufacturers.

9 The liner should be poured in one continuous operation after the whitemetal has been thoroughly mixed and stirred and the surface fluxed and cleared of dross.

10 It is desirable to cool the steel quickly if possible as this minimises shrinkage porosity and improves adhesion. This should be done carefully by water-spraying from the back and bottom of the shell.

11 After the bearing has cooled, dismantle the fixture, check the lining for porosity and remove any excess metal.

12 Check the quality of the bond between liner and shell by tapping the shell with a small hammer. It should give a clear, ringing sound. A cracked sound indicates that the bond is poor.

13 Clamp the two halves of the bearing together and set them up in the lathe ready for machining the inside diameter. Check the journal size and the manufacturer's recommendations and machine the bore of the bearing with the best possible finish.

14 Remount the bearing after applying a coating of Prussian blue (bearing blue) to the journal. Turn the shaft over a couple of times and then inspect the bearing's surfaces. The blue should transfer reasonably evenly to the bearing's surfaces. If it is only present on small areas this means that the bearing is riding on high spots. See Fig. 6-19.

HIGH SPOTS ON BEARING SURFACE

Fig. 6-19 Testing bearing fit with bearing blue.

Scrape the bearing surface using proper scraping tools to remove the high spots. Reassemble and recheck. Repeat the process until the fit is satisfactory.

15 If oil grooves need to be machined, ensure that the edges are chamfered afterwards.

The procedure for relining solid, whitemetal-lined bushings is identical to the above, except that the set-up for pouring is slightly different. See Fig. 6-20.

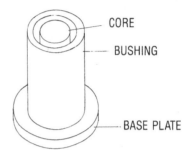

CORE

BUSHING

BASE PLATE

Fig. 6-20 Pouring set-up for relining solid, whitemetal-lined bushings.

Start-up

The following procedures should be adopted during the start-up of new or reconditioned plain bearings.

1 Before the shaft is run, check the alignment with bearings in position. Bluing of the journal will indicate whether the shaft is running true in the bearing. If the wear spots are at one end of one of the bearing halves and at the opposite end of the other as shown in Fig. 6-21 then misalignment is indicated.

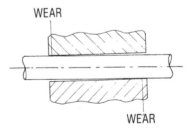

WEAR

WEAR

Fig. 6-21 Bluing indicates misalignment.

Alignment should be corrected by shimming the bearing housing.

2 Once the alignment is true, pre-lubricate the bearing surfaces before final assembly to ensure they do not start up dry.

3 Run the shaft and check for noise and vibration.

Check bearing temperatures every half hour for the first two or three hours of operation. If temperatures are excessive, shut down the machine and inspect the bearing surfaces for wear-in characteristics. Further alignment or scraping may be required before re-starting.

4 Once the operation of the bearing is satisfactory, ensure that it is regularly supplied with the correct lubricant.

FAILURE PATTERNS

Before replacing a damaged bearing it is important to know the precise cause of failure so that the problem can be properly corrected, otherwise similar failures may keep occurring due to the same cause.

Bearing problems and failures can be analysed in terms of effects or symptoms or in terms of causes. From the point of view of fault detection it is useful to look first of all at symptoms and then to trace the possible causes. Every failure is different in some respects so it is extremely difficult to establish definite rules about causes. The same symptom may result from many different causes and systematic troubleshooting techniques are necessary to eliminate all but the real cause. However, it is possible to establish some general rules about the relationship between symptoms and causes that can serve as a guide in the troubleshooting process.

There are two levels at which the symptoms of bearing failure can be detected. The first is what might be called the level of **external** symptoms which are exhibited while the bearing is in operation. The second can be called the level of **internal** symptoms which can only be observed when the bearing is shut down, dismantled and inspected.

Operating symptoms

When the condition of a bearing begins to deteriorate, whatever the cause, then it may exhibit one or all of the following symptoms:

Overheating

The temperature of a plain bearing during operation is a vital indication of its condition. A properly lubricated bearing in good condition should not generate excessive temperatures and should run cool enough to touch. The most efficient way to check bearing temperature is to use a portable sensing device and to measure the normal running temperature of the bearing once it has been run in. Routine checks at regular intervals will then show up any sudden or dramatic increases from the base level.

When using a contact pyrometer or thermocouple to measure temperature it is important to ensure that good contact is achieved at a position that truly reflects the temperature of the bearing.

Vibration

A plain bearing in good condition should run very smoothly, even more so than a rolling element bearing. As clearances increase, some vibration may develop which will tend to accelerate the deterioration of the bearing. As with temperature, hand contact is usually sufficient to establish whether unnecessary vibration is present. It is advisable where possible, and especially for critical equipment items, to establish a normal operating level of vibration using a vibration monitor (see Chapter 10) and to make routine checks to establish when a deviation occurs. As with temperature, a dramatic change in operating characteristics indicates a need for further investigation.

Noise

A plain bearing should not produce high noise levels if it is in good condition. If the noise level increases it may be due either to misalignment or excessive wear. Foreign particles trapped in the bearing may also cause noisy operation. In a reciprocating mechanism the big and little end bearings produce a 'knocking' sound when wear becomes excessive. Noise levels can be measured independently although vibration monitoring is more commonly used. Technicians should develop an 'ear' for the sound of machinery so that changes in both level and characteristics can be detected.

Seizure

Total seizure of a plain bearing is, of course, a symptom which cannot be ignored. If the three characteristics referred to above are carefully monitored then seizure should not occur. Seizure is usually a result of excessive temperature build-up due to lack of lubrication and if temperatures are routinely checked then seizure should be avoided. The danger in allowing undetected deterioration to lead to ultimate seizure of a bearing is that such a sudden catastrophic failure is likely to also cause severe damage to the housing and other machine components. The excessive temperature build-up can result in fire and the sudden shutdown of the machine may cause severe or even disastrous process problems.

Symptoms often occur simultaneously and the presence of one often leads to the others appearing. This means that cross-checks can be made to establish whether a problem exists. For example, if it is suspected that the noise level is increasing, then a check on operating temperature may confirm that the condition of the bearing is deteriorating.

Symptoms found on inspection

The operating symptoms alone rarely, if ever, provide sufficient evidence for the exact cause of failure to be determined. It may be possible, once the external symptoms of failure are detected, to attribute them to the general cause of 'wear' and then merely to proceed to replace the bearing and re-start the equipment. This is a short-sighted approach however, and runs a strong risk of a premature repetition of the same problem. In order to be more precise about the cause of failure, it is necessary to dismantle the bearing and inspect the bearing surfaces to establish what internal symptoms exist. These can generally be identified according to one of the following categories.

Wear

Any two surfaces that are in constant sliding contact, even though properly lubricated, will eventually wear out. This process will, in most cases, be apparent due to a general loss of metal across the bearing surface. The bearing surface may still be reasonably good and the wear only clearly evident by the increased clearances established by a dimension check. Wear occurs due to a combination of the processes of adhesion and abrasion described in Chapter 5.

Mobil

Fig. 6-22 A typical example of general wear.

Fatigue

Fatigue is caused by cyclical stressing due to fluctuating loads and although not as common in plain bearings as in rolling element bearings it can often occur when load or speed becomes excessive.

Mobil

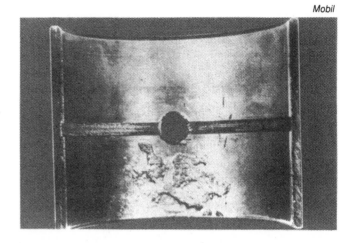

Fig. 6-23 A typical pattern of fatigue failure.

Galling

Sometimes the adhesion due to cold welding of two surfaces in contact becomes so great that grain displacement of the bearing material takes place leaving pits or craters. This is usually associated with high loads which cause excessive pressure between points of contact.

Mobil

Fig. 6-24 Galling of the bearing surface.

Scoring

When excessive amounts of dirt or large dirt particles are present due to contamination of the lubricant or

Mobil

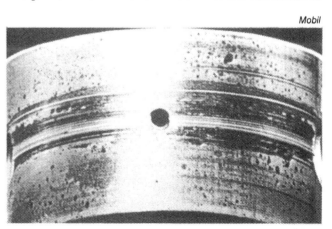

Fig. 6-25 Scoring of the bearing surface.

because of abrasive wear then scoring and erosion of the bearing surface will result.

Wiping

When there is insufficient clearance in the bearing or overheating occurs due to inadequate lubrication then a superficial melting of the bearing surface and a flow of material occurs.

Mobil

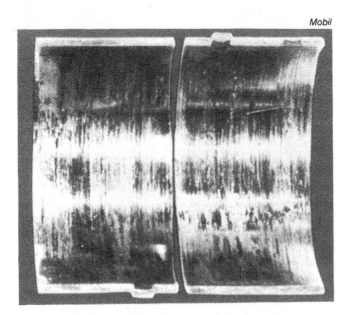

Fig. 6-26 Wiping of the bearing surface.

Fretting

Fretting is most likely to occur in plain bearings between the liner and the housing when the interference fit is inadequate and slight vibratory movement occurs. A characteristic fine brown debris of oxidation material is the usual evidence of fretting or false brinelling as it is sometimes called.

Glacier Metal

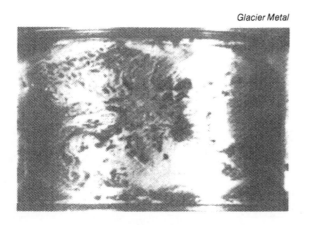

Fig. 6-27 Fretting between liner and shell.

Fretting may also occur between the bearing surfaces while the shaft is stationary due to vibrations transmitted from external sources.

Glacier Metal

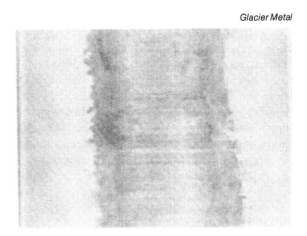

Fig. 6-28 Fretting due to external vibration.

Pitting

Local loss of metal in the form of pits or deep craters can result from a number of causes. This can sometimes be due to stray electric currents that arc across the bearing and cause pitting of both the surface of the bearing and the journal.

Mobil

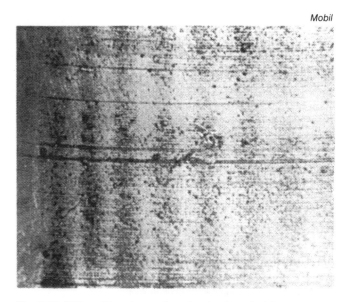

Fig. 6-29 Pitting of bearing surface due to stray electric current.

Corrosion

Etching and surface discoloration are all evidence of corrosion, and a typical example is shown in Fig. 6-30.

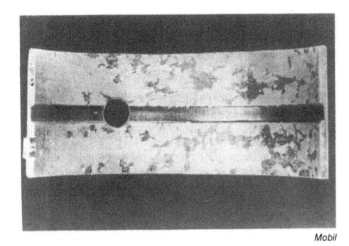

Mobil

Fig. 6-30 Typical evidence of corrosion.

Causes of failure

In theory there are an almost infinite number of possible causes of bearing failure but in practice it has been found that most failures can be attributed to one of several quite specific causes. In order of significance, these are as follows:

Dirt

The presence of dirt is the single most destructive cause of plain bearing failure. Large dirt particles which become embedded in the bearing surface can cause excessive wear and reduce the life of the bearing and journal. An excessive quantity of fine dirt particles will cause scoring and rapid wear. Dirt particles can also become trapped between the bearing liner and the housing. When the particles are large this can cause distortion of the liner, reduction of clearances and an impairment of heat transfer. Over-heating and, in some cases, fatigue failure are the likely results.

Dirt particles are most likely to become entrapped during the assembly procedure and the technician must take all possible care to ensure that this does not occur. Contaminated lubricant may also be a source of dirt and recirculating systems should be properly filtered and the filter changed or cleaned at regular intervals. In a crankcase the failure of one bearing may result in contamination of the lubricant by particles of bearing material which may then endanger other bearings. If this condition is suspected then more frequent lubricant changes may be necessary to protect the other bearings.

Inadequate lubrication

Oil starvation is a common cause of bearing failure and may occur for various reasons. The period immediately after overhaul is always critical and it is recommended that the lubrication system be primed to ensure lubricant is present on initial start-up. The lubrication system should be kept under close observation for the period after start-up to ensure that pressures and temperatures are established according to design.

During normal operation, several factors may affect the supply of oil. Leaks, oil pump failure, blocked filters and blocked oil passages are all potential problems. Care must be taken when installing bearing liners to ensure that the oil hole is in line with the oil passage in the housing. It should also be recognised that once a bearing becomes badly worn the hydrodynamic wedge cannot be effectively established and the wear rate will further increase due to inadequate lubrication.

Bearing failure may also occur due to use of an unsuitable lubricant due either to error or poor selection. Lubricant suppliers and equipment manufacturers should be consulted if there is reason for concern about the suitability of the lubricant used. Lubricants should be stored properly and clearly marked so that the likelihood of error is reduced.

Improper assembly

As well as ensuring that cleanliness is observed during the assembly procedure, it is also important to ensure that the bearing is assembled correctly in all other respects. The main considerations are that the bearing should be fitted square and secure in its housing and that it does not get physically damaged during assembly.

Failure to ensure proper assembly may result in uneven or rapid wear and overheating.

Misalignment

If the journal is not properly aligned in the bearing uneven bearing wear and possible damage to the journal will result. The alignment procedures described earlier in this chapter should be followed carefully. The alignment of shaft couplings should also be checked to ensure that misalignment there does not affect the true running of the journal in the bearing. (See Chapter 8.)

Overload

Equipment overloads can occur for many different reasons and are particularly likely to occur when extra output is required. The most common symptoms of overload are overheating and fatigue. When these are observed a review of operating conditions is advisable in order to establish whether any overloading has occurred.

Increased speeds, feed rates, levels, operating temperatures and pressures, etc. can all create overload conditions which lead to machine elements such as bearings having to perform above design limits.

Moisture

The presence of water due to condensation or lack of effective sealing may cause corrosion of the bearing surfaces and the formation of rust.

Lubricant breakdown

Breakdown of the lubricant, because of the formation of organic acids due to oxidation or the presence of specific elements in the oil such as sulphur compounds, may lead to corrosion of the bearing surfaces. An analysis of the lubricant after it has been in service may be necessary to establish whether deterioration has taken place. A review of

the oil replacement period may be sufficient to avoid recurring problems. This should be done in consultation with the lubricant supplier.

Summary of the common symptoms and causes of plain bearing failure

Symptoms		Causes
Operating	Inspection	
Noise	Wear	Dirt
Overheating	Fatigue	Lack of lubricant
Vibration	Galling	Improper assembly
Seizure	Scoring	Misalignment
	Wiping	Overload
	Fretting	Moisture
	Pitting	Lubricant breakdown
	Corrosion	

6.2 ROLLING ELEMENT BEARINGS

PRINCIPLES OF OPERATION

Rolling element bearings, or anti-friction bearings as they are often called, differ from plain bearings in that they incorporate rolling elements, either balls or rollers, which are held between two raceways as shown in Fig. 6-31. A soft metal cage or retainer separates the rolling elements and ensures that they are evenly spaced, but does not carry any load. As a result of the relatively small area of contact between the rolling elements and the races, the frictional resistance to relative motion is comparatively low.

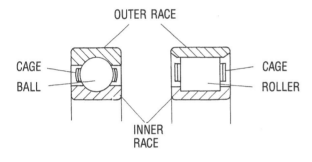

Fig. 6-31 Typical ball and roller bearings.

Because the relative motion between the moving elements is one of rolling rather than sliding, the material requirements are quite different from plain bearings. Instead of requiring bearing materials to be soft in comparison with the journal, both the rolling elements and the raceways of rolling element bearings are made of specially hardened steel. The harder the elements, the smaller the indentation and deformation of the surface and hence the lower the frictional resistance. However, although sliding action between the two moving elements is much reduced, in comparison with plain bearings, it cannot be eliminated entirely, and therefore lubrication is just as critical to rolling element bearings as to plain bearings.

Rolling element bearings are manufactured to very precise standards of accuracy to ensure good performance and long life. Tolerances are held to a thousandth or in some cases a tenth of a thousandth part of a millimetre in order to minimise run-out and to ensure the proper radial and axial clearances required for smooth operation.

The choice of balls or rollers as the rolling elements depends on the operating conditions and the following considerations.

Ball bearings
can operate at higher speeds without overheating,
are less expensive for lighter loads,
have lower frictional resistance at light loads,
are available in a wider range of sizes.
Roller bearings
can carry heavier loads,
are less expensive for heavier loads and larger sizes,
are superior under shock or impact loading,
provide greater rigidity.

The important physical difference between balls and rollers that gives rise to this difference in performance is the variation in the area of contact between the rolling elements and the raceway. The ball has a small area of contact which resembles point contact depending on how much deformation of the ball and raceway occurs. The roller, by comparison, has a greater area of contact which resembles line contact. The difference is shown in Fig. 6-32.

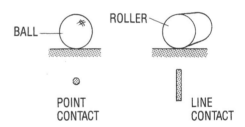

Fig. 6-32 Difference in area of contact between ball and roller bearings.

The greater area of contact of the roller makes it better able to carry heavy loads and withstand impact, but tends to increase frictional resistance at low loads. Ball bearings generate lower resistance at low loads but deform more significantly at high loads which causes frictional resistance to increase. Ball bearings can run at higher speeds due to the smaller area of contact.

TYPES OF BEARINGS

Journal bearings

	Type	Radial load	Axial load	Speed	Misalignment tolerance	Axial location
1	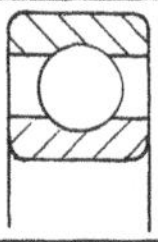 Single row deep groove ball	very good	good	excellent	good	both directions
2	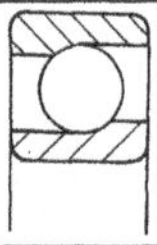 Single row angular contact ball	very good	very good	good	fair	one direction
3	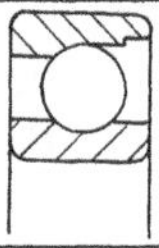 Magneto	fair	fair	very good	poor	one direction both directions when used in pairs
4	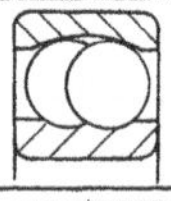 Self aligning ball	good	very poor	very good	excellent	both directions
5	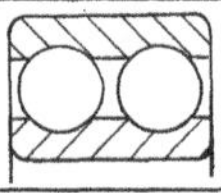 Double row deep groove ball	very good	good	excellent	fair	both directions
6	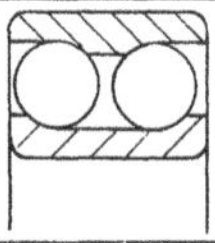 Double row angular contact ball	very good	very good	good	fair	both directions
7	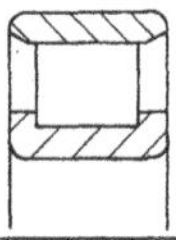 Roller	excellent	very poor	very good	poor	—
8	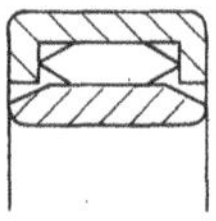 Needle roller	excellent	very poor	fair	very poor	—
9	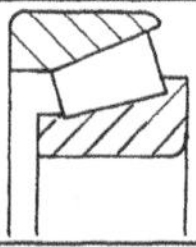 Tapered roller	excellent	very good	good	poor	one direction
10	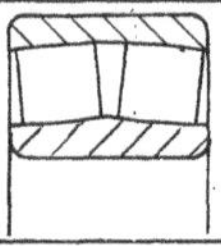 Spherical roller	very good	good	fair	excellent	both directions

Fig. 6-33 Types of rolling element journal bearings.

Thrust bearings

	Type	Radial load	Axial load	Speed	Misalignment tolerance	Axial location
11	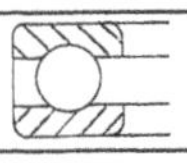 Single row ball	—	very good	fair	poor	one direction
12	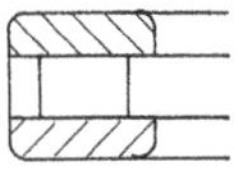 Cylindrical roller	—	excellent	poor	very poor	one direction
13	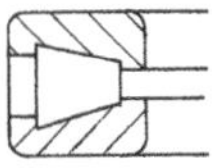 Tapered roller	—	excellent	poor	very poor	one direction
14	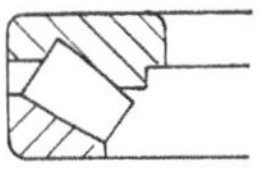 Spherical roller	poor	very good	poor	very good	one direction
15	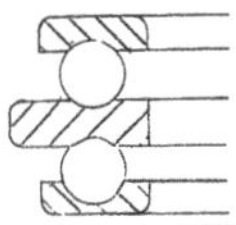 Double row ball	—	good	poor	poor	both directions

Fig. 6-34 The types of rolling element thrust bearings.

Linear bearings

The construction of slider or linear motion bearings is slightly different from that of journal and thrust bearings.

Ladder bearings

These consist of two hardened steel plates that are separated by a series of balls or rollers held in a cage as shown in Fig. 6-35. They may be used either vertically or horizontally.

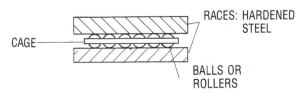

Fig. 6-35 Ladder bearing.

Recirculating ball and roller bearings

In these units the rolling elements are only in contact with the way or slide for a specific distance, then they leave the load area, drop into a return channel and return to the opposite end of the assembly to be fed back into the loaded area. The roller version can carry greater load than the ball but tends to have a shorter life.

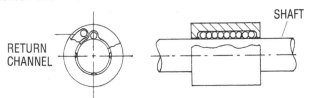

Fig. 6-36 Recirculating ball bearing.

Recirculating roller-chain bearings

An endless chain assembly of rollers is mounted on a solid bearing race attached to a carriage. The rollers are often shaped to fit a ground shaft or way. As the carriage or slide moves, the bearing races roll on the concave rollers which are in contact with the shaft.

Single mounted units have little resistance to side loads and hence two units are often mounted in a V configuration.

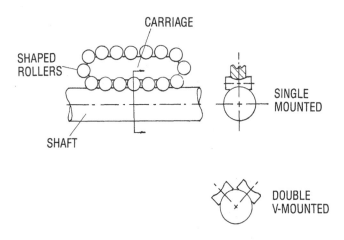

Fig. 6-37 Recirculating roller-chain bearing.

BEARING ASSEMBLY DESIGN

Shaft and housing fits

As a general rule a rolling element bearing should be installed so that its rotating ring is an interference fit and the stationary ring is a slip or push fit. An interference fit is one which requires a press or a driver or, if the fit is especially tight, may require hot mounting, whereas a slip or push fit can be slid into place by hand. Sometimes if loads and speeds are high, both races may be fitted with an interference fit.

   The fit required for a particular bearing should be established by consulting the manufacturer or the standards of the ISO (International Standards Organisation).

Thermal expansion

When a shaft is mounted on rolling element bearings some provision may have to be made for thermal expansion of the shaft in the axial direction while at the same time ensuring that all elements of the machine are maintained in the correct relative positions.

   In the common case where the shaft rotates and is held by more than one bearing it is normal to have one bearing held in the housing and the other floating and able to accommodate axial movement of the shaft due to expansion or contraction. A typical arrangement is shown in Fig. 6-38.

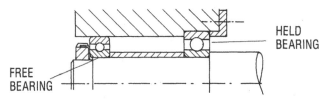

Fig. 6-38 Typical arrangement of free and held bearings.

Types of mountings

There are several standard arrangements that are used to fix a bearing to a shaft so that its axial position is maintained.

Shaft nut and locking washer

This is the most common arrangement and can be used for both single and double row radial bearings. It employs a specially designed nut and a tab washer as shown in Fig. 6-39.

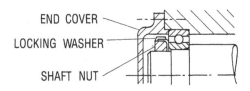

Fig. 6-39 Shaft nut and locking washer arrangement.

Slotted nut and pin

Sometimes a castle nut or specially designed slotted nut may be used in conjunction with a split pin or cotter pin in order to provide adjustment for the inner cones of tapered roller bearings as shown in Fig. 6-40.

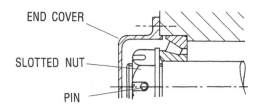

Fig. 6-40 Slotted nut and pin.

End plate and locking bolts

This type of device is often used when the bearing is mounted on the end of the shaft as shown in Fig. 6-41.

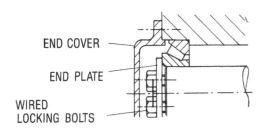

Fig. 6-41 End plate and locking bolts.

End cap and shims

This arrangement is also used for tapered roller bearings or for angular contact bearings where pre-load has to be set by adjusting shims.

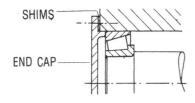

Fig. 6-42 End cap and shims.

Snapring

When it is not suitable for the housing to contain a shoulder then a snapring can be used to locate the bearing in the housing as shown in Fig. 6-43.

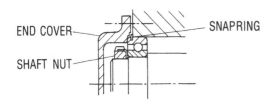

Fig. 6-43 Snapring.

Adapter sleeve

Spherical roller bearings are usually made with a tapered bore and are often mounted as 'floating' bearings on an adapter sleeve as shown in Fig. 6-44.

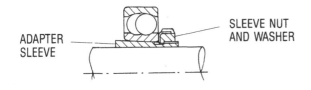

Fig. 6-44 Adapter sleeve.

Withdrawal sleeve

A withdrawal sleeve of the type shown in Fig. 6-45 is also used with self-aligning bearings for ease of dismounting. In this case the bearing must be mounted against a shoulder.

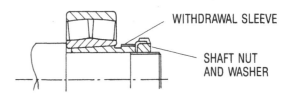

Fig. 6-45 Withdrawal sleeve.

MAINTENANCE PRACTICES

General

Although there are specific maintenance practices associated with particular types of rolling element bearings there are a number of general recommendations that apply to all types. These should be kept in mind whenever bearing maintenance is carried out.

- As with plain bearings, cleanliness is vital to the satisfactory operation of rolling element bearings. Bearings should be kept free of dust and dirt particles and be kept dry and protected at all times.
- Bearings should be handled with clean, dry hands or with clean gloves.
- It is important when mounting and dismounting bearings that the correct tools are used and that they are in good condition. Makeshift arrangements or badly worn tools are likely to lead to damage to the bearing or bearing assembly.
- Bearings should be wrapped in oil-proof paper when not in use.
- Only clean solvents and flushing oils should be used for bearing cleaning.
- Before installing a rolling element bearing both the housing and shaft should be carefully inspected for burrs, nicks and scratches that may interfere with the fitting of the bearing.
- Bearings that are dry and unlubricated or have not been cleaned should not be spun. If compressed air is used to clean a bearing care should be taken to avoid spinning the bearing.
- Cotton waste or dirty rags should not be used to clean bearings, but only clean, lint free rags.
- The slushing compound used to protect a bearing

in storage need not be removed if it is petroleum-based unless it has gone hard or become contaminated.

- The force applied when mounting or dismounting a bearing should always be applied to the ring with the interference fit and should never be applied in such a way that the force is transmitted through the rolling elements.
- Never strike a bearing directly with a hammer or mallet or with a soft metal drift.
- Remember that all sealed and shielded bearings must be mounted cold.
- Bearings should never be heated with a naked flame.

Tools and equipment

The maintenance of rolling element bearings requires specialised tools and equipment and without these many tasks are extremely difficult. The following items should be considered essential:

Mounting dollies and sleeves

When mounting bearings which are a press fit in the housing or on a shaft it is important that pressure is applied evenly around the bearing ring. If the bearing is cocked due to uneven force then surfaces can be damaged and the bearing ring distorted. A set of mounting dollies and sleeves will ensure that bearings are driven home evenly. These can be purchased direct from the bearing manufacturers or can be made up to suit a particular application.

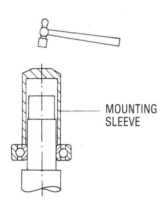

MOUNTING SLEEVE

Fig. 6-46 A typical example of a mounting sleeve.

Hammers

Hammers should be made from steel or a soft material and should be free from burrs. Copper or synthetic resin are suitable materials for soft hammers but lead and tin should not be used. Wooden mallets should not be used because of splinters.

Drifts

Only steel drifts should be used and they should only be used for bending or straightening tabs on locking washers or for driving shaft nuts.

Arbor press

An arbor press ensures that bearings are driven evenly especially when the fit is tight. It should be used in preference to other methods where practicable.

Pullers

Bearing pullers are essential for dismounting and there are several types available.

a.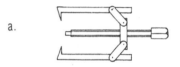

b.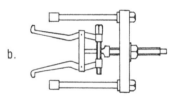

Fig. 6-47 Typical examples of bearing pullers.

Pullers should be maintained in good condition so that they operate freely and the claws are free of burrs.

Lifting gear

When handling large, heavy bearings it is important to have suitable lifting gear available such as tongs or slings.

Hook and impact spanners

Hook and impact spanners are needed for use with adapter sleeve nuts, withdrawal nuts and shaft nuts.

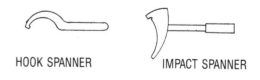

HOOK SPANNER   IMPACT SPANNER

Fig. 6-48 Hook and impact spanners.

Induction heater or oil bath

Modern induction heaters, such as the type shown in Fig. 6-49, are much cleaner and easier to use than an oil bath when bearings have to be heated for a shrink fit.

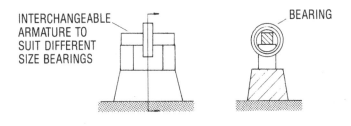

Fig. 6-49 Typical induction heater.

When induction heaters are not available, the traditional oil bath should be used.

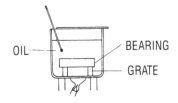

Fig. 6-50 Oil bath.

Gauges

A set of various types of gauges is necesssary for measuring diametral clearances, housing bores and shaft diameters. An ordinary set of feeler gauges is sufficient for measuring clearances but a special set of bore gauges may be required for measuring the bore of housings. Alternatively, standard inside and outside micrometers can be used for measuring diameters.

The above tools represent the minimum required for bearing maintenance. Where bearings are mounted on tapered shafts or on adapter or withdrawal sleeves some hydraulic equipment may also be required.

Oil injection equipment

Where shafts have been provided with oil ducts, hydraulic pressure supplied by an oil injection pump can be used for mounting and dismounting both straight bore and tapered bore bearings. A typical situation is shown in Fig. 6-51.

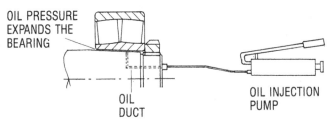

Fig. 6-51 Mounting by oil injection.

Hydraulic nuts

These tools consist of a nut that incorporates a groove in which an annular piston moves when oil pressure is built up behind it. Hydraulic nuts can be used for both mounting and dismounting and employ the same oil injection equipment referred to above.

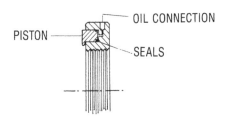

Fig. 6-52 A typical hydraulic nut.

As well as the above equipment, there are a number of standard items which may be required from time to time and should be readily available. These may include:
    straight edge
    marking blue
    dial gauge
    plumb line
    solvent
    clean rags, etc.
Tools should be kept clean, and in good condition. Bearings are precision items and tools and equipment should therefore be maintained, handled and stored accordingly.

Mounting procedures

Preliminary steps

Before starting the mounting procedure for any bearing, the following points should be considered:
- Check the manufacturer's drawing of the bearing arrangement and ensure that it is clearly understood. Establish which fit is the interference fit.
- Check that the necessary tools and equipment are available after determining the mounting procedure to be used.

- Select a suitable working environment that is clean and adequately lit.
- If an old bearing is to be remounted make sure that it has been properly cleaned and lightly coated with lubricant. Protect the bearing in greaseproof paper until ready for fitting.
- New bearings should be kept wrapped until ready for fitting. Check that the bearing is the correct type and size according to the manufacturer's recommendations.
- Check that the shaft and bearing housing are clean and free from burrs and other surface damage.
- Check the shaft and housing dimensions to ensure that they are correct according to the manufacturer's drawings.
- Check shoulders and abuttments for run-out, especially in the case of thrust bearings. Run-out in the thrust face mounting will cause rapid wear and should be checked with a dial gauge.

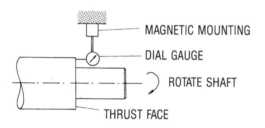

Fig. 6-53 Checking run-out in the thrust face mounting.

Pre-lubrication

Normally, rolling element bearings are not lubricated until after they are mounted, although there may be exceptions, especially when the bearings are inaccessible. For cold mounted bearings, however, a light lubrication of the bearing seat and the shaft journal and housing will assist in the assembly procedure. It is also good practice where shaft nuts and adapter sleeve nuts are used for drive-up, that the threads be lightly lubricated so that they create minimum resistance.

Selection of mounting method

The method to be used for mounting a bearing will depend on the type and size of bearing and on the mounting arrangement. The manufacturer's instructions or service manual should be followed where available.

Bearings with a bore of 100mm (4″) or less can usually be cold mounted, whereas larger bearings need to be heated. For bearings with tapered bores, only very large bearings need to be heated.

Cylindrical bore bearings need to be mounted mechanically whereas tapered bore bearings can be mounted using hydraulic mounting tools.

The following methods can be considered as standard for mounting rolling element bearings. If doubt exists as to which method should be used then equipment and bearing manufacturers should be consulted.

Arbor press

An arbor press can be used for small bearings with cylindrical bore. A sleeve should be used between the bearing and the press which has flat, parallel are burr-free end faces. The sleeve should bear on the bearing ring with the interference fit.

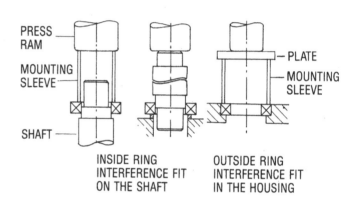

Fig. 6-54 Typical arrangements using an arbor press.

Hammer and dolly

For cold mounting of all cylindrical bore bearings a suitable dolly or sleeve can be used to drive the bearing on to its mounting. The sleeve must bear on the bearing face with the interference fit and care should be taken to ensure that the bearing is driven on smoothly and does not cock over.

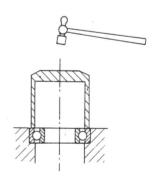

Fig. 6-55 Typical example of the use of a hammer and dolly.

Hook and impact spanners

For small and medium sized bearings (i.e. up to 200 mm (8″) bore) with tapered bore, either a hook or impact spanner can be used with a drive nut to drive the bearing on to the tapered shaft or on to an adapter sleeve as shown in Fig. 6-56.

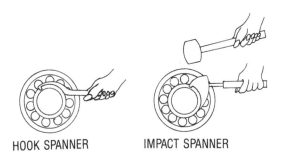

HOOK SPANNER   IMPACT SPANNER

Fig. 6-56 Using hook and impact spanners.

When a drive nut is used in this manner the face of the nut facing the bearing should be coated with a dry lubricant such as molybdenum disulphide and the surface of the shaft or sleeve coated with a light oil.

Hydraulic nut

A hydraulic nut can be used for bearings with tapered bores as shown in Fig. 6-57.

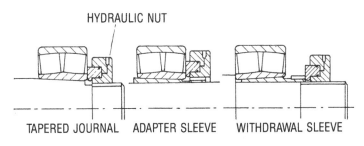

HYDRAULIC NUT

TAPERED JOURNAL   ADAPTER SLEEVE   WITHDRAWAL SLEEVE

Fig. 6-57 Using a hydraulic nut.

When oil is pumped into the nut the annular piston forces the bearing on to the tapered seat until it reaches a shoulder or the required amount of axial drive-up (p.46).

The hydraulic nut should be used in conjunction with a suitable oil pump and oil of viscosity recommended by the manufacturer.

Oil injection

Oil can be injected between the bearing inner face and the shaft for a tapered bore bearing to expand the bearing and reduce the friction and make it easier to drive the bearing on to the seat as shown in Fig. 6-51.

In order to be able to use this method the shaft must have been specially machined with an oilway and oil grooves to distribute the oil between the bearing and the shaft.

The viscosity of the oil and the pressure developed by the pump should be determined with the supplier of the equipment.

Sometimes withdrawal sleeves are also specially machined with oil ducts and can be used with oil injection as shown in Fig. 6-58.

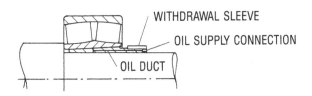

WITHDRAWAL SLEEVE

OIL SUPPLY CONNECTION

OIL DUCT

Fig. 6-58 Withdrawal sleeve machined with oil ducts.

Oil injection plus hydraulic nut

For large bearings on tapered bore shafts or with suitably machined withdrawal sleeves, oil injection can be used in conjunction with a hydraulic nut.

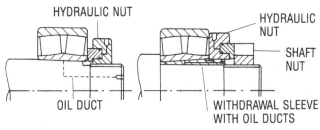

HYDRAULIC NUT

HYDRAULIC NUT

SHAFT NUT

OIL DUCT

WITHDRAWAL SLEEVE WITH OIL DUCTS

Fig. 6-59 Using oil injection in conjunction with a hydraulic nut.

Hot mounting

For bearings with cylindrical bores and for large bearings with tapered bores it may be necessary to heat the bearing. The bearing should be heated to around 80°C to 90°C above the shaft temperature but never to more than 120°C (250°F). The heating apparatus should be close to the equipment and the hot bearing should be pushed home quickly and smoothly before it cools down or jams. An induction heater of the type shown in Fig. 6-49 should be used where possible.

If an oil bath is used, the bearing should be heated with the oil and not dropped into hot oil and should be kept off the bottom of the bath by a grate to prevent distortion as shown in Fig. 6-50.

If these two methods are not available then a hot plate or an oven may be used.

Whatever method is used a careful check must be kept on the temperature of the bearing by using a surface thermometer.

When a bearing is an interference fit in the housing and it is impossible to heat up the housing it may be necessary to cool the bearing by submerging it in a bath of alcohol cooled by dry ice or a cryogenic liquid.

Special considerations

Adjustment of tapered roller bearings

Tapered roller bearings must be set up either with clearance or with a certain amount of preload, depending on the manufacturer's instructions. The simplest way to do this is to draw up the shaft nut or end plate until there is no play in the bearings and drag becomes noticeable. If end play is required then the shaft nut can be backed off and if preload is required then it can be pulled up harder. The adjustment can be measured by mounting a dial gauge against a shaft shoulder or gear face.

In an arrangement with shims, such as the one shown in Fig. 6-60, once the play has been taken up the gap between the end plate and shaft end can be measured with a feeler gauge and either the end play added or the preload subtracted in order to establish the correct thickness of shim pack.

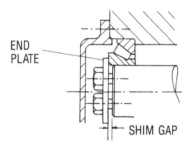

Fig. 6-60 Adjusting the shim pack for tapered roller bearings.

Where clearance is required it is always wise to take a final check of end play with a dial gauge against a shaft shoulder as shown in Fig. 6-61.

If a shaft locking nut arrangement is involved then a final check of clearance will be needed after the locking nut has been tightened because this will force the shaft nut towards the bearing by an amount equal to the clearance in the threads.

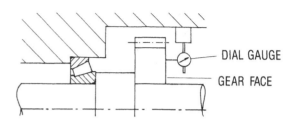

Fig. 6-61 Checking end play.

Axial drive-up

For a bearing mounted on a tapered bore, the degree of interference between the inner ring and the shaft depends on how far the bearing is driven up the tapered shaft. This dimension is known as axial drive-up. In small spherical roller bearings, where clearances cannot be measured, the axial drive-up is used as a measure of the interference fit.

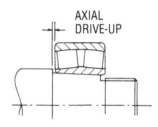

Fig. 6-62 Axial drive-up used as a measure of interference fit.

Manufacturers' information should be consulted for correct values of axial drive-up for different bearings.

Measurement of clearances in spherical roller bearings

The internal clearance in a spherical roller bearing is greater than in a ball bearing and can be measured with feeler gauges. Because spherical roller bearings are mounted on tapered seats, as the bearing is pushed on to the taper the inner ring expands and reduces the internal clearance in the bearing. Hence the final clearance is a direct function of the interference fit on the shaft. Measurement of the clearance is useful in giving an indication of the shaft fit. This can be accomplished as follows:

• Measure the unmounted radial clearance by standing the bearing on a clean surface and rotating the inner ring backwards and forwards to seat the rollers properly in the outer ring. Then use a feeler gauge to establish the clearance between the uppermost roller and the outer ring. Do this by inserting the gauge between two top rollers and then rolling the roller under the blade. Increase the feeler gauge thickness until the roller traps the blade and cannot be withdrawn.

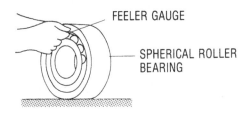

Fig. 6-63 Using a feeler gauge to establish clearance.

- When the bearing is mounted on the tapered shaft or sleeve, the clearance must be measured at the bottom instead of the top. As before, the outer ring should be rotated a few times to ensure that the rollers are properly seated and a feeler gauge inserted between the rollers and the bottom of the outer ring. See Fig. 6-64.

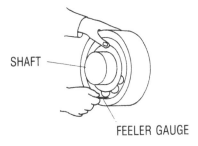

Fig. 6-64 Using a feeler gauge to establish clearance.

- With self-aligning ball bearings the clearances are too small to measure with a feeler gauge. Normal practice is to tighten up the shaft nut and to check the clearance by swivelling and rotating the outer ring. When the clearances are correct the ring should rotate freely but there should be some resistance to swivelling.

Dismounting procedures

The procedure used for dismounting a bearing will depend on the way in which the bearing is mounted and whether the interference fit is on the shaft or in the housing. If the bearing is to be re-used then its relative position should be marked before it is dismounted i.e. which side is 'up' and which way does

it face. Whether the bearing is to be re-used or not, care should be taken not to damage the bearing during the dismounting process so that the evidence of failure is not disturbed and can be used to establish cause of failure. The dismounting force, like the mounting force, should always be applied to the ring with the interference fit. Care should also be taken not to damage the surface of the shaft or housing.

The following methods are commonly used.

Interference fit on the shaft

The most common method for dismounting bearings with either a cylindrical bore or a tapered bore is by the use of a puller such as the one shown in Fig. 6-47a.

Bearings mounted on shafts that have been machined with oil ducts can be dismounted using the oil injection method.

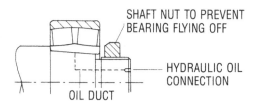

Fig. 6-65 Using the oil injection method.

Interference fit in the housing

When the bearing is an interference fit in the housing then the bearing may have to be hammered off using a dolly or a soft metal drift.

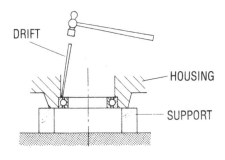

Fig. 6-66 Hammering off the bearing.

Because of the tendency of soft metal drifts to chip they should never be used for mounting. When dismounted in this way bearings should be washed carefully before being re-used.

For bearings where the inner ring can be swivelled

an inside puller can be used, such as the one shown in Fig 6-47b.

Bearings on adapter sleeves

A hammer and dolly can be used either to drive the bearing off the sleeve or to drive the sleeve from under the bearing, depending on which way round the sleeve is mounted.

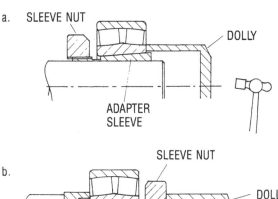

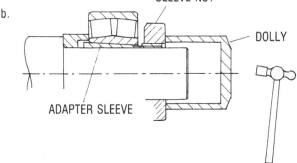

Fig. 6-67 Using a hammer and dolly to dismount a bearing from an adapter sleeve.

The alternative is to use a hydraulic nut as shown in Fig. 6-68.

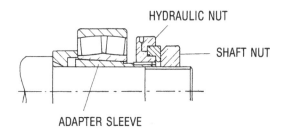

Fig. 6-68 Using a hydraulic nut for dismounting.

Bearings on withdrawal sleeves

For small and medium sized bearings a hook or impact spanner can be used to drive up a withdrawal nut to force out the sleeve as shown in Fig. 6-69.

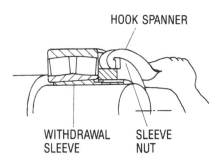

Fig. 6-69 Forcing out the sleeve with a hook spanner.

To make the process easier, the threads and faces of the withdrawal nut should be lubricated with molybdenum disulphide.

For large bearings the use of a hydraulic nut is recommended as shown below in Fig. 6-70.

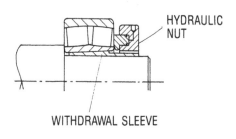

Fig. 6-70 Using a hydraulic nut to remove a withdrawal sleeve.

If the withdrawal sleeve is machined with oil ducts then oil injection can be used in conjunction with the withdrawal nut and an impact spanner, as shown in Fig. 6-71.

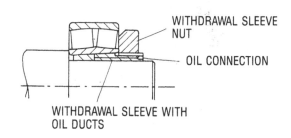

Fig. 6-71 Oil injection used in conjunction with a withdrawal nut.

FAILURE PATTERNS

As in the case of plain bearings, it is important to accurately determine the cause of failure before a rolling element bearing is replaced or repaired. This requires careful analysis of the symptoms of malfunction and careful inspection of the failed or damaged components. Generally speaking, malfunction and failure of rolling element bearings are caused by much the same problems as affect plain bearings. The symptoms of failure can be classified into those noticed during operation and those found on inspection. The evidence of failure differs to some extent from that shown by plain bearings because of the physical difference in construction.

Operating symptoms

The external symptoms of failure are the same for rolling element bearings as for plain bearings.

Overheating

As the condition of a rolling element bearing deteriorates, the frictional resistance to motion increases and the operating temperature of the bearing begins to rise. Unlike plain bearings, however, where almost any change in condition tends to cause an increase in operating temperatures, only certain causes lead to overheating in rolling element bearings. The two main causes are inadequate lubrication and loss of internal clearance. Because of the sliding contact between the surfaces of a plain bearing almost any change in surface condition will increase the frictional resistance to motion. The rolling contact in a rolling element bearing reacts rather differently.

The temperature of a rolling element bearing can be measured in the same way as a plain bearing and once again it is deviation from the normal running condition that provides a warning of malfunction. A rise of 10°C above normal running temperature is generally considered as cause for alarm with a maximum allowable temperature for a standard bearing being around 125°C (260°F).

Vibration

Unlike plain bearings, some vibration of rolling element bearings is normal due to the action of the rolling elements themselves. With rolling element bearings it is particularly important that a **signature pattern** (see Chapter 10) is established either by vibration analysis or by direct observation. Deviation from that pattern can then be monitored to detect change in condition of the bearing. The acceptable limits for measured vibration vary with the size and type of machine and recommended values are given in Chapter 10. If vibration levels are assessed by direct observation then experience is required to determine the severity of the problem.

An understanding of the principles of operation of the bearing and the material characteristics should make it obvious, even to the inexperienced, when vibration levels are becoming unacceptable.

Vibration may develop in a bearing because of defects in the bearing itself, or it may be transmitted to the bearing from other parts of the machine or other adjacent equipment. Vibration produces rapid and repeated blows on bearing surfaces and contributes, sooner or later, to fatigue failure of the bearing material. It can also affect the properties of grease and damage bearing seals and hence lead to loss of lubricant.

When vibration is transmitted to a stationary bearing it can cause fretting corrosion, described below.

Noise

As is the case with vibration, rolling element bearings generate more noise under normal running conditions than do plain bearings. The normal noise pattern may vary due to changes in operating speeds and loads and, as with vibration patterns, a signature should be established from which deviations can be observed.

Although noise can be measured with an acoustic monitoring device (see Chapter 10) it is more usual for direct observations to be made by ear. Hence it is important for maintenance personnel to become familiar with the characteristic noises produced by different types of bearings and the changes produced by different types of defects. For example, dirt tends to produce a crackling sound, damage to the raceways causes a high-pitched whine, and loss of lubricant causes a squeaking sound. The most common aids used in the detection of bearing noise remain the screwdriver and the stethoscope and valuable information about bearing condition can be determined this way.

Seizure

If noise, vibration and overheating are carefully monitored then complete seizure should rarely occur. When it does, it may indicate that overheating has reached the point where the bearing metals have fused and the usual cause of this is lack of lubrication. It is also possible for a very badly worn bearing to seize due to displacement of the rolling elements. The failure of the bearing cage or retainer can also cause the rolling elements to jam.

Symptoms found on inspection

Rolling element bearings also develop symptoms of malfunction that can only be observed when the bearing is dismantled. Careful observation of the bearing surfaces and other evidence is essential if the precise cause of failure is to be established. The only exception here is sealed bearings which are usually discarded and replaced without investigation.

The following symptoms are commonly found as evidence of rolling element bearing malfunction or failure.

Wear

Rolling element bearings are not as susceptible to general wear from adhesion and abrasion as plain bearings, because of the rolling rather than sliding motion of the surfaces, but the presence of foreign material will cause surface damage in the form of scratching and scoring as shown in Fig. 6-72.

SKF

Fig. 6-72 Surface damage caused by the presence of foreign material.

Fatigue

The rolling action of balls and rollers sets up an intermittent transfer of load from one raceway to another and generates cyclical stresses in the loaded area of the raceways. The continuous deflection of the metal causes cracks to develop which lead to flaking of the surfaces.

SKF

Fig. 6-73 Surface flaking.

As this condition worsens, the metal breaks down further and grain displacement takes place. This condition is known as **spalling.**

SKF

Fig. 6-74 Grain displacement due to spalling.

Fretting

Fretting corrosion or false brinelling is most likely to occur when a rolling element bearing is dry and stationary. It is due to slight, almost imperceptible motion between the contacting surfaces of the rolling elements and raceways. Such motion may be transmitted to the bearing from an external source or may come from the machine itself. Fretting produces a characteristic red-brown dust at the interface.

SKF

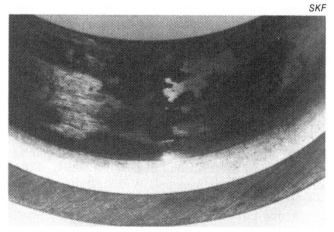

Fig. 6-75 A typical example of fretting.

Fretting corrosion is most likely to occur when a machine is in transit or on stand-by.

Brinelling

True brinelling, which false brinelling or fretting is

said to resemble, consists of indentations in the raceways made by the rolling elements when the bearing is subjected to shock loading.

Mobil

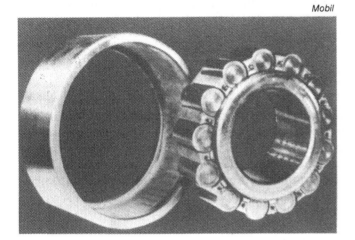

Fig. 6-76 True brinelling.

Pitting and fluting

A pitted or a fluted effect as shown in Fig. 6-77 may appear as a result of stray electric currents arcing across the contact surfaces.

SKF

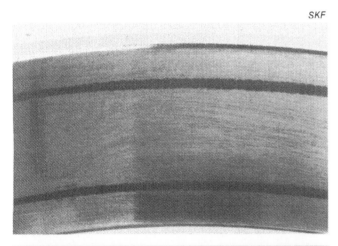

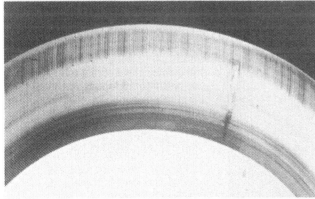

SKF

Fig. 6-77 Pitting and fluting.

It is usual for the pitting to develop first and if no action is taken to prevent further damage this deteriorates further into a fluted or washboard effect.

Smearing and galling

If the rolling elements fail to rotate properly and rub or slide over the raceways then smearing may occur due to the surface of the raceway 'picking up' on the rolling element as shown in Fig. 6-78.

SKF

Fig. 6-78 Smearing of bearing surface.

In extreme cases this may cause grain displacement and is then referred to as galling which can be identified by the craters produced as shown in Fig. 6-79.

Mobil

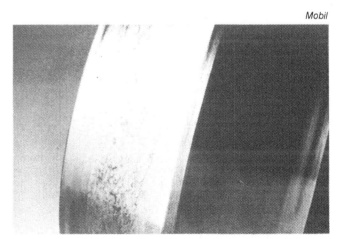

Fig. 6-79 Evidence of galling.

Corrosion

Corrosion of bearing surfaces may be evident as pitting or spalling of bearing raceways. Rust and other corrision products may be present.

Mobil

Fig. 6-80 A typical example of corrosion.

Wear patterns

The pattern or load zone produced on the internal surfaces of a ball bearing can be an important clue to the cause of failure. Although other types of bearings also generate specific loading patterns they are much more difficult to analyse than ball bearings.

In order to benefit from the study of load zones it is necessary to differentiate between normal and abnormal patterns. The normal pattern to be

expected from a bearing will depend on the way the load is carried and whether it is the inner or the outer ring that rotates. Fig. 6-81 shows examples of the normal and abnormal load patterns that can be expected to develop under the loading conditions stated.

Causes of failure

The same problems that cause malfunction and failure of plain bearings also cause failure of rolling element bearings. Rolling element bearings can also be affected by factors that do not usually affect plain bearings due to their different configuration. Thus the symptoms described above can result from one or more of the following:

Dirt

As with plain bearings, the presence of dirt or other foreign material in the bearing will cause excessive wear and premature failure of a rolling element bearing. Cleanliness during assembly, constant

NORMAL PATTERNS

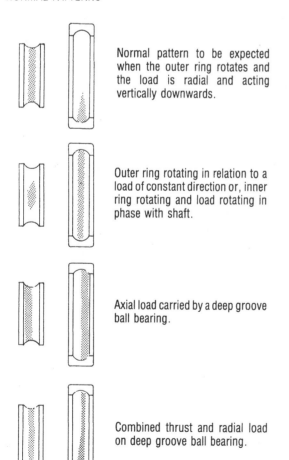

Normal pattern to be expected when the outer ring rotates and the load is radial and acting vertically downwards.

Outer ring rotating in relation to a load of constant direction or, inner ring rotating and load rotating in phase with shaft.

Axial load carried by a deep groove ball bearing.

Combined thrust and radial load on deep groove ball bearing.

ABNORMAL PATTERNS

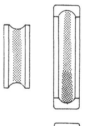

If the interference fit between the inner ring and shaft is too tight the bearing will be pre-loaded and both rings will be loaded with a wide pattern over 360°.

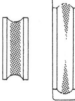

Another form of internal pre-load can be produced by a distorted or out of round housing which causes two or more load zones to appear on the outer ring.

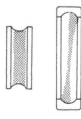

Misalignment will produce load zones that are not parallel to the ball groove. This diagram shows the pattern to be expected when the outer ring is misaligned in relation to the shaft.

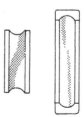

This diagram shows the pattern to be expected when the inner ring is misaligned relative to the housing.

Fig. 6-81 Typical loading patterns for ball bearings.

filtering of lubricants and well maintained seals will help to avoid this problem.

Inadequate lubrication

The oil supply system is as critical for rolling element bearings as for plain bearings and all the recommendations for maintaining an adequate supply of the correct lubricant apply. Failure of the system will lead to rapid deterioration in bearing condition due to smearing and galling, and the likely seizure of the bearing. It is generally recognised that inadequate lubrication is the single most common cause of rolling element bearing failure.

Improper assembly

If a rolling element bearing is not correctly assembled then rapid wear will result from excessive or insufficient pre-loading, misalignment or failure to secure the inner and outer rings correctly. Abnormal wear patterns such as those described in the previous section will result and also the likelihood of fatigue failure is increased.

Misalignment

For rolling element bearings that do not have inherent self-aligning features the alignment of shaft and housing can be critical to the operation of the bearing. Cylindrical roller bearings are the most susceptible to run-out and there are limits to the tolerable run-out for all other types except self-aligning ball bearings and spherical roller bearings. Wear patterns should be studied carefully in order to establish evidence of misalignment and squareness of the housing and shaft alignment should be closely checked if doubt exists.

Overload

If a rolling element bearing is subjected to loads in excess of those for which it was designed, then its life will be shortened and early fatigue failure will occur. The reduction in life span can generally be expected to be proportional to the cube of the load increase. This means if the load were doubled then bearing life would be reduced to approximately one-eighth of normal. Overloading can also occur due to excessive speed which leads to overheating of the bearing. High temperatures may affect the properties of the lubricant and cause centrifugal throw-out of oil and grease. Excessive wear will occur as the effectiveness of the lubricant deteriorates, and at very high speeds fatigue life may be reduced due to the increased loading which results from high centrifugal forces acting on the rolling elements.

Moisture

The presence of water in a bearing, due to leakage into the housing or condensation of moisture due to temperature changes, can cause considerable damage from corrosion. Water may also find its way into the bearing as a result of improper washing and drying techniques used during inspection.

In an oil lubricated bearing, the presence of significant quantities of water will cause emulsification of the lubricant.

Lubricant breakdown

Oxidation of the lubricant may lead to the formation of acidic compounds that may cause corrosion of bearing surfaces. Sometimes lubricant additives may be incompatible with bearing materials and also lead to corrosive attack.

Shock loading

Rolling element bearings are more susceptible to shock loading than are plain bearings because of the point or line contact of the rolling elements. Shock loads cause indentation of the raceways and result in the brinelling effect discussed in the previous section.

Electric currents

Stray electric currents, even at very low voltages, can arc across the lubricant film and lead to pitting and fluting. This is more likely to occur with electrical machinery but can also occur in other machines due to currents that build up statically and short circuit to ground through the bearing. This type of problem may also occur due to the incorrect positioning of earthing leads when welding is being carried out on a machine.

External vibrations

Vibrations transmitted from other parts of a machine, or from another machine altogether, can cause damage to a rolling element bearing whether it is running or stationary. If a bearing is in operation and is subjected to vibrations from an external source then its fatigue life is likely to be reduced depending on the severity of the vibrations. External vibrations transmitted to a stationary bearing may produce fretting corrosion.

Summary of the symptoms and causes of rolling element bearing failure

Symptoms		Causes
Operating	Inspection	
Overheating	Wear	Dirt
Vibration	Fatigue (spalling)	Inadequate lubrication
Noise	Fretting	Improper assembly
Seizure	Brinelling	Misalignment
	Pitting and fluting	Overload
	Smearing and galling	Moisture
	Corrosion	Lubricant breakdown
	Abnormal wear zones	Shock loading
		Electric currents
		External vibration

POWER TRANSMISSION

As was discussed in Chapter 2, most rotating machines operate in sets which include a driver and driven machine at least and often include a transmission machine such as a gear box as well.

There are various ways in which the output shaft of one machine can be linked to the input shaft of another so that transmission of power can take place.

7-1 V-belt drives

Although various other types of belt drives are also used, V-belts are the most common. However, many of the maintenance considerations that apply to V-belts are also relevant to other types of belt drive such as flat belts and timing belts.

PRINCIPLES OF OPERATION

V-belts are normally used to transfer power between two shafts whose axes are parallel and some distance apart.

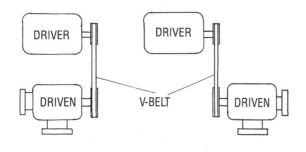

Fig. 7-1 Typical V-belt arrangements.

The belt is mounted on pulleys that are attached to the driving and driven shafts and the drive relies on friction between the belt and the pulleys for its operation. The belt sits in the groove of the pulley and makes contact with the sides of the groove as shown in Fig. 7-2.

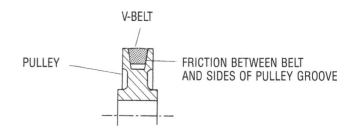

Fig. 7-2 V-belt in pulley.

In order to be able to transmit power, the belt must be under tension so that it is forced down into the groove. The belt is squeezed and friction develops between the sides of the belt and the sides of the groove. The depth of the groove is always greater

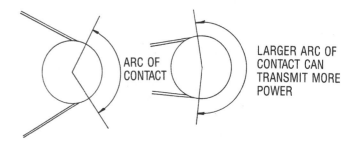

Fig. 7-3 Relationship between power and arc of contact.

than the thickness of the belt, however, and the belt should never bottom in the groove. The operation of the belt and its ability to transmit power depend on the size of the friction force and the arc of contact of the belt. The greater the arc of contact the more power the belt can transmit. (Fig. 7-3)

As well as performing its primary function of transmitting power, a V-belt can be used to change the speed of the driver output and hence the torque transmitted to the driven unit. There are three basic alternatives as shown in Fig. 7-4.

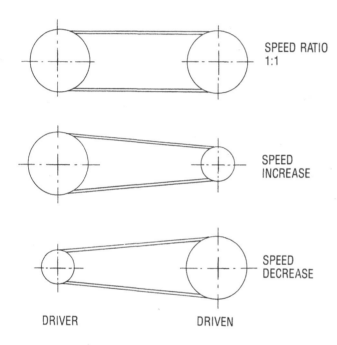

Fig. 7-4 Alternative arrangements for V-belt drives.

The speed ratio between the two pulleys of a belt drive can be calculated from a simple formula.

$$\frac{\text{driven speed}}{\text{(RPM)}} = \frac{\text{driver pulley diameter (mm)}}{\text{driven pulley diameter (mm)}} \times \frac{\text{driver speed}}{\text{(RPM)}}$$

It is generally accepted that V-belt drives are limited to belt speeds between 300 and 3000 metres per minute (1000–10 000 feet per minute). If required to operate at higher speeds then dynamic balancing of the pulleys becomes increasingly important.

TYPES AND ARRANGEMENTS

Single V-belt

The most common type is the single belt arrangement whose operation has been described above. In addition to being used to transmit power between parallel shafts, the single belt can also be used for quarter turn drives and angle drives as shown in Fig. 7-5.

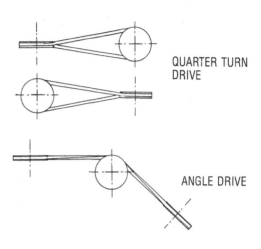

Fig. 7-5 V-belts can be used for quarter turn and angle drives.

Multiple V-belt

In order to increase the capacity of the drive an arrangement which uses several belts mounted on multi-grooved pulleys is often used.

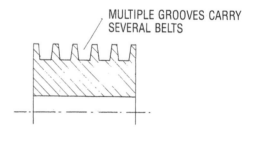

Fig. 7-6 Multi-grooved pulley.

Banded V-belts

In order to overcome the tendency of belts to whip, twist or jump off, a banded V-belt, in which the V-sections are vulcanised to a common band can be used.

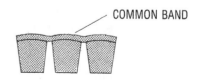

Fig. 7-7 A banded V-belt.

Belt tension

V-belts are often tensioned by means of an idler pulley for the following reasons.
- If the relative position of the shafts cannot be adjusted then an idler pulley can be used to assist installation of the belt.
- If the drive is subject to varying loads then a spring loaded idler can provide automatic adjustment of belt tension.
- The inclusion of an idler pulley can help to increase the arc of contact and hence the power transmission capacity of the drive.

It is generally recommended that idler pulleys be mounted on the slack side of a belt, as shown in Fig. 7-8, and positioned close to the drive pulley.

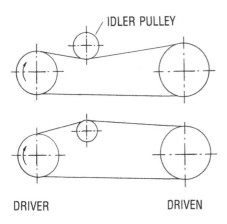

Fig. 7-8 Mounting of idler pulleys.

MAINTENANCE PRACTICES

The following general points should be taken into account in the maintenance of V-belts:
- The operation of V-belts depends largely on the condition and correct positioning of the pulleys.
- V-belt pulleys should be kept clean, free of oil and grease and free from damage and wear.
- Pulleys should be installed parallel and in line with each other.
- Belt tension should be adjusted according to the manufacturer's recommendations.
- V-belts should never be forced or levered on to the pulley as shown in Fig. 7-9.
- Multiple V-belts should be correctly matched to ensure the load is evenly distributed. As well as being the same size it is preferable that all belts are supplied by the same manufacturer.
- Multiple V-belts should be changed as a set. A single new belt will be shorter than the worn, stretched belts and will tend to carry more than its

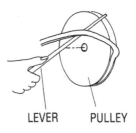

Fig. 7-9 Incorrect method of installing V-belt on pulley.

fair share of load. It is therefore likely to fail prematurely.
- No dressings of any kind should be applied to a V-belt.
- V-belts should be stored in a clean, dry place and should not be exposed to heat or direct sunlight.
- Do not hang V-belts on nails or small pegs while in storage. Store flat if possible.

Alignment

The correct alignment of the shafts and pulleys is vital to the operation of a V-belt drive. Misalignment of pulleys can occur in several ways.

The first step in aligning the pulleys is to check that the two shafts are level and parallel. This should be done by using a spirit level on the exposed shafts to

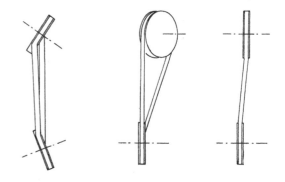

Fig. 7-10 Misalignment of pulleys.

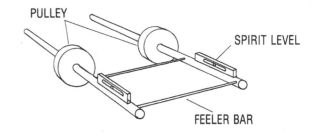

Fig. 7-11 Using a spirit level and feeler bar.

check for level, and then by using a feeler bar or gauge to check the distance between the shafts on both sides of the pulleys. (Fig. 7-11)

Once the shafts are parallel then the pulleys can be brought into line by using a straight edge across the faces as shown in Fig. 7-12.

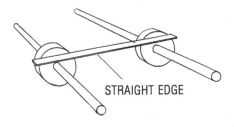

Fig. 7-12 Pulleys brought into line using a straight edge across the faces.

Reposition either of the pulleys until they are properly aligned. If either shaft is subject to end float make sure that it is in its normal running position when the alignment is checked. Rotate the shafts and check the alignment in several positions before it is finally accepted.

Belt tension adjustment

It is important that V-belts run with the amount of tension recommended by the manufacturer. If tension is insufficient then the belt will slip and overheating and wear will result. If the belt is too tight it will also cause overheating as well as damage to bearings.

There are a number of ways in which belt tension can be checked. The most common is to depress the belt and measure the deflection using a ruler and straight edge as shown in Fig. 7-13.

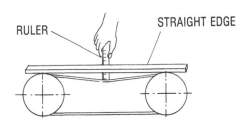

Fig. 7-13 Checking belt tension by depressing the belt and measuring the deflection.

If this method is not considered sufficiently accurate then a spring balance can be used to deflect the belt against a specified pull as shown in Fig. 7-14.

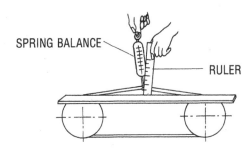

Fig. 7-14 Using a spring balance.

A third alternative is to measure the elongation of the belt under tension. This is done by marking a defined length of the belt and then remeasuring between the same two marks when the belt is under tension as shown in Fig. 7-15.

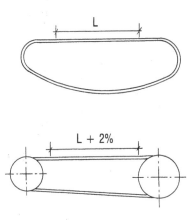

Fig. 7-15 Measuring the elongation of the belt under tension.

For normal drives the elongation of the belt should be around 2%. This may increase for high speed drives and should be checked against manufacturer's recommendations.

Installation of V-belt drives

The following procedure should be adopted when installing V-belts.

1 Check the condition of the pulleys and check for wobble.

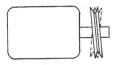

Fig. 7-16 Pulley wobble.

2 Align the pulleys according to the procedures outlined above.
3 Reduce the centre distance of the shafts by adjusting the position of the motor. If an idler pulley is used release the tension on the idler so that the belt can be fitted.
4 Fit the belt or belts taking care not to damage them in the process.
5 Adjust the belt tension as recommended above.
6 Run the unit for a short period, say 10 minutes, to allow the belt to seat correctly in the pulley grooves.
7 Readjust the tension.
8 Recheck the tension after 24 hours of operation.
9 Ensure that the drive is protected with a suitable guard at all times during operation. The guard should be provided with ventilation and secure against removal by unauthorised personnel.

FAILURE PATTERNS

Like most machine elements, V-belts do not have an unlimited life and will eventually wear out. Their normal life expectancy will depend on the operating conditions, speed and loading. When belts fail prematurely it is important to determine the precise cause of failure so that corrective action can be taken and performance improved for future operation. As with other machine elements an investigation of failure patterns can be divided into an analysis of symptoms and an analysis of causes.

Symptoms fall into two categories, those apparent during operation and those visible on shutdown of the machine and inspection of the drive.

Operating symptoms

The following conditions can be considered to be evidence of malfunction.

Belt slippage

Any tendency of a V-belt to slip will lead to rapid wear and premature failure. The common causes of belt slippage are insufficient tension, drive overload and the presence of oil or grease on the belt.

Belt squeal

Squealing often accompanies belt slippage and is also caused by overload and insufficient tension. It may also occur when the arc of contact between belt and pulley is insufficient.

Belt ticking or slapping

When the operation of a belt drive is accompanied by a ticking or a slapping sound this is often evidence that some form of mechanical interference is taking

place. This may be due to poorly aligned guards or contact with other machine parts.

Belt whipping

If a V-belt starts to whip it is likely to jump out of the pulley groove or to roll over and become damaged. Whipping may be the result of the drive centres being too far apart or due to wobbling pulleys. Sometimes a pulsating load will also cause belt whip, in which case the suitability of the drive should be reviewed.

Belts turned over

If the cords in the belt are broken during installation by levering the belt on to the pulley then the belt will stretch excessively and lose strength. Lack of tension may also allow a belt to roll over in the pulley groove. The effect of impulse loads and whipping may also cause the belt to roll over and this may be overcome by installing a spring loaded idler pulley. Once a belt has turned over it will be damaged and should be automatically discarded and replaced.

Belt breakage

If a V-belt breaks immediate action is clearly required. Breakage may occur due to overloading in the form of shock loads or heavy starting loads but if the belt has been properly selected this should not occur. A belt that has been weakened by being levered on to the pulley is very likely to break prematurely. The presence of foreign objects or material may also damage the belt sufficiently for it to break.

Symptoms found on inspection

When the drive is shut down and examined the condition of belts and pulleys will provide evidence from which the cause of failure may be determined. As for all machine elements, every failure is different in some way. However, the following conditions are commonly-found symptoms of V-belt malfunction.

Wear

A properly aligned and tensioned belt will wear along the sides and will eventually need to be

Gates Rubber

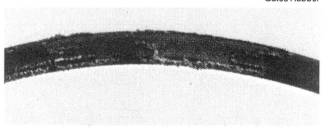

Fig. 7-17 A typical example of normal wear.

replaced. If wear is rapid and leads to premature failure this may be due to misalignment or the presence of dust or other abrasives.

Uneven wear, as shown in Fig. 7-18, may be the result of either misalignment or damage to the pulley grooves.

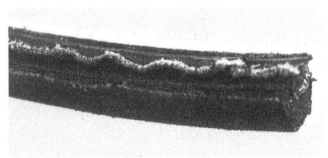

Gates Rubber

Fig. 7-20 Fraying of the belt.

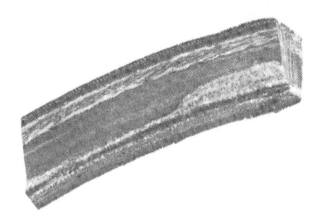

Gates Rubber

Fig. 7-18 Example of uneven wear.

Cracking

When hardening and cracking appear on the underside of the belt this is usually caused by excessive heat build-up. This may be caused by poor ventilation or by slippage.

Gates Rubber

Fig. 7-19 Hardening and cracking of the belt.

Fraying

Any tendency for the belt to fray along the edges or for the surface to tear and rupture, as shown in Fig. 7-20, is usually evidence that some mechanical interference is taking place or that the pulleys are worn or damaged in some way.

Stretching

If a belt stretches beyond the adjustment range of the tightener then the chances are that the internal cords are broken and the belt should be replaced.

Swelling

If the belt material becomes swollen or spongy it is likely that it has been exposed to oil, grease or other chemicals.

Burns

If a belt shows evidence of burning in one particular area this may indicate either that the belt has slipped during start-up or that the driven unit has jammed or stalled causing the belt to burn when the drive has run on.

Gates Rubber

Fig. 7-21 Evidence of burning in one particular area.

Worn sheaves

In addition to belt damage the pulley grooves may show evidence of wear. This may appear on the sides of the groove (Fig. 7-22), or as a shiny surface on the bottom of the groove (Fig. 7-23) which indicates that the belt has been bottoming. A pulley which shows excessive wear of this type should be replaced and new belts installed.

WEAR ON SIDES
OF GROOVES

Fig. 7-22 Dishing of the sides of the groove.

WEAR ON BOTTOM
OF GROOVE

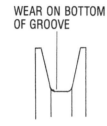

Fig. 7-23 Evidence that the belt has been bottoming.

Causes of failure

The common causes of the types of V-belt failure listed above can be summarised as follows:

Misalignment

As with most mechanisms good alignment is critical. Misalignment of pulleys will cause rapid wear and damage to the belt surface, although alignment tolerances for V-belts are not as stringent as for other drives. The manufacturer's instructions should be consulted to establish the appropriate limits for a particular machine.

Incorrect tension

Insufficient tension will cause the belt to slip and may also lead to belt breakage due to the grab-slip effect. Over-tensioning will increase the wear rate and shorten belt life and may also cause overloading of shaft bearings.

Interference

Any contact between the belt and another part of the machine or the belt guard will cause rapid belt damage in the form of fraying or excessive wear of the belt surface.

Foreign material

Belts and pulley grooves should be kept clean and free of dirt, grit and other contaminants which may cause accelerated wear or even belt breakage. Oil or other chemicals will attack the belt and cause deterioration of the material. A well-constructed belt guard should help to exclude foreign material.

Damaged pulleys

Pulley grooves should be free from nicks, burrs, chips and other damage that may affect belts.

Overloading

The life of a belt drive will be reduced if it is subjected to loads or speeds beyond its design capacity. Excessive wear and belt breakage is likely to occur if loads and speeds are too high.

Overheating

Excessive heat is an enemy of V-belts and can lead to their rapid deterioration. Guards should be made of mesh to allow adequate ventilation, and belt slippage should be corrected as quickly as possible before the heat build-up affects the belt material.

Summary of the common symptoms and causes of V-belt failure

Symptoms		Causes
Operating	Inspection	
Slipping	Wear	Misalignment
Squealing	Cracking	Incorrect tension
Ticking	Fraying	Interference
Whipping	Stretching	Foreign material
Turn-over	Swelling	Overloading
Breakage	Burns	Overheating
	Worn sheaves	Damaged sheaves

7.2 CHAIN DRIVES

Chains and sprockets provide a positive form of drive which does not slip and they can therefore be used where synchronisation of motion is important. There are various types of chains available, the most common being the roller chain dealt with here. Many of the maintenance practices described also apply to other types of chains such as silent chain, rollerless chain, etc.

PRINCIPLES OF OPERATION

Chains and sprockets fulfil the same basic function as belts and pulleys in transferring power between two parallel shafts. Instead of relying on friction, a chain drive is a positive drive in which the links of the chain engage with specially formed teeth on the sprocket.

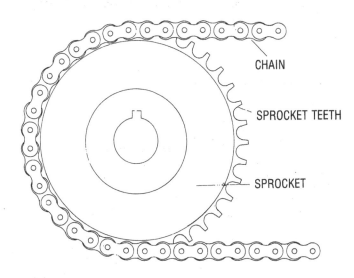

Fig. 7-24 Chain and sprocket.

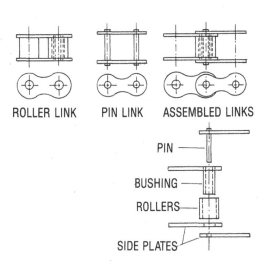

Fig. 7-25 Standard roller chain.

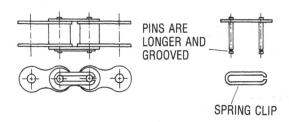

Fig. 7-26 Special joining link.

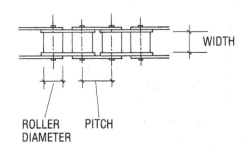

Fig. 7-27 Pitch, width and roller diameter are the critical dimensions of roller chain.

Standard roller chain is made up of alternate roller links and pin links.

The pitch of the chain is determined by the length of the side plates, and the bushings and pins are press-fitted into the side plates. The pins of a special joining link may be longer and grooved to take spring clips as shown in Fig. 7-26.

The rollers are free to rotate on the bushings, and this reduces the rubbing action between the chain and the sprocket as the chain links roll on to the sprocket and thus avoid excessive wear from sliding friction. Because each roller and bushing functions like a plain journal bearing, lubrication is essential to the operation of chain drives.

The critical dimensions by which roller chain is identified are the pitch, width and roller diameter.

Chain drive sprockets have teeth cut around the periphery, like a gear, and are specified by the pitch circle diameter, width and number of teeth. They are usually manufactured with an integral hub as shown in Fig. 7-28.

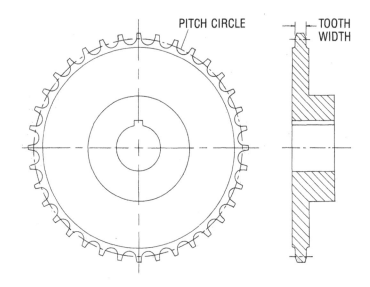

Fig. 7-28 Chain drive sprockets are usually manufactured with an integral hub.

It is normal practice to design chain drives in such a way that the number of chain pitches and the number of sprocket teeth ensure that the same link does not contact the same tooth each revolution. If there are an even number of pitches on the chain there must be an uneven number of teeth on the sprocket and vice versa. This helps to reduce uneven wear.

Chain drives are more sensitive to misalignment than belt drives and must be properly tensioned. They are generally suitable for speeds up to 1350 metres per minute (4500 feet per minute).

The speed of the driven sprocket in relation to the speed of the driver can be determined by using a simple formula based on the number of teeth on the driver and driven sprockets:

$$\begin{array}{l}\text{speed of driven}\\\text{(RPM)}\end{array} = \begin{array}{l}\text{speed of driver}\\\text{(RPM)}\end{array} \times \dfrac{\text{no. of teeth on driver}}{\text{no. of teeth on driven}}$$

TYPES AND ARRANGEMENTS

Standard roller chain is available in single and multi-strand form, and the number of strands required will depend on the power to be transmitted. Double pitch chains are also available. They are cheaper, and are suitable for light loads and low speeds.

Chain drives are used most commonly as horizontal drives and any slack in the chain, resulting from wear, should accumulate on the lower strand as shown in Fig. 7-30.

Vertical drives should be arranged so that accumulated slack falls into the driven sprocket rather than away from it, to prevent misengagement.

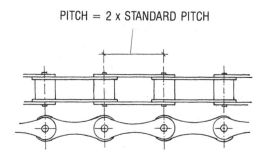

Fig. 7-29 Double pitch chain.

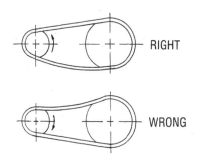

Fig. 7-30 In a horizontal drive, slack should accumulate on the lower strand.

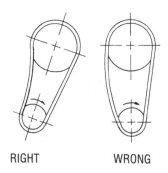

Fig. 7-31 Accumulated slack in a vertical drive should fall into, rather than away from, the driven sprocket.

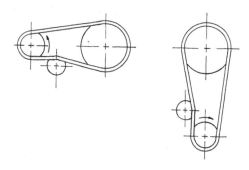

Fig. 7-32 Using a chain tensioner.

Where chain tensioners are used they should be used on the side of the chain where the slack is expected to accummulate. (Fig. 7-32)

MAINTENANCE PRACTICES

The general points to be taken into account in the maintenance of chain drives are:
- As with belt drives, alignment and proper tensioning are critical to the operation of the drive.
- Chain drives should be kept clean and protected from dirt and should be provided with an adequate supply of a suitable lubricant.
- New links should not be installed in chains that have been significantly lengthened by wear.
- New chains should not be installed on badly worn sprockets. Sprockets may be reversed on the shaft to extend their life if necessary.
- New chains should be stored in protective wrappings until ready for use, and protected from excessive heat and moisture.
- Chain drives, like belt drives, should be properly guarded and protected from interference.

Alignment

The alignment of chain sprockets is critical to the operation of the drive and the procedure should be carried out with care. As for V-belt pulleys, the two shafts are first checked for level and parallel alignment using a spirit level and feeler bars or gauges.

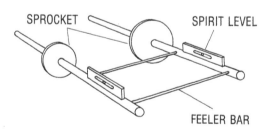

Fig. 7-33 Checking for level and parallel alignment of the shafts.

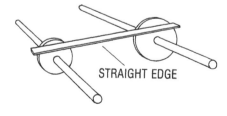

Fig. 7-34 Checking alignment of the sprocket faces.

A straight edge can then be used to check alignment of the sprocket faces as shown in Fig. 7-34.

If either of the shafts is subject to end float then align the sprockets with the shaft in its normal running position.

If the centre distance between the shafts is too great for a straight edge a taut piano wire may be used instead. Rotate the sprockets and check the alignment in several positions.

Slack adjustment

Unlike belt drives, chain drives do not require initial tension but should merely be adjusted to take up the slack. If a chain is too tight it will bind on the sprockets and wear rapidly. If it is too loose it will tend to whip, which will cause vibration and reduced life.

For chains installed on units with adjustable centres, the slack should be adjusted to approximately 2% of the centre distance. In other words, if the centre distance is one metre then the slack should be about 20mm. This can be adjusted by pulling the chain taut on one side and measuring the slack on the other using a ruler and straight edge.

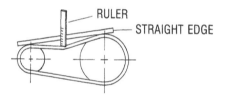

Fig. 7-35 Measuring the slack in a unit with adjustable centres.

For drives on fixed centres, an idler sprocket or spring loaded tightener of some kind will automatically take up the slack.

It is important to recognise that chains stretch during operation and drives with adjustable centres may need to be readjusted from time to time. The so-called stretch is not due to any physical deformation of the chain links but due to wear in the rollers, bushings and pins which must be compensated for.

Installation of chain drives

The following procedure is recommended when installing a chain drive.
1 Check the condition of the sprockets and make sure they are clean and free from damage.
2 Align the sprockets as recommended above.
3 Loosen the tighteners so that the chain will fit over the sprockets.

4 Remove the chain from its wrapping and bring the ends together over one of the sprockets. Use the sprocket teeth to hold the chain and install the final link.

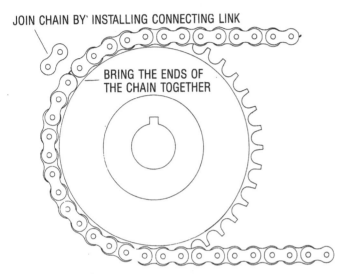

JOIN CHAIN BY INSTALLING CONNECTING LINK

BRING THE ENDS OF THE CHAIN TOGETHER

Fig. 7-36 Using the sprocket teeth to hold the chain.

5 Fit the side plate and spring clips or whatever device is used to secure the pins.
6 Take up the slack in the drive as outlined above.
7 Lubricate the chain according to the manufacturer's instructions.
8 Start up the machine and check that the chain runs true, without excessive noise and without binding or whipping.
9 Ensure that the lubrication system is working properly.
10 Install a suitably designed guard to prevent interference with the drive. (Do not attempt to do this while the drive is in operation.)

Chain removal

If the drive adjustment allows the chain to be removed without being broken by the removal of a link then use this method. If the chain then needs to be shortened by removing a link a chain detacher should be used to hold the chain while the link is removed. (Fig. 7-37)

Chain detaching tools are also available to drive the link pins out. (Fig. 7-38)

Cleaning

Whatever the operating conditions it is advisable to remove the chain and clean it from time to time. The debris of normal wear and gummed lubricant will cause wear to the pins and bushings if it is not

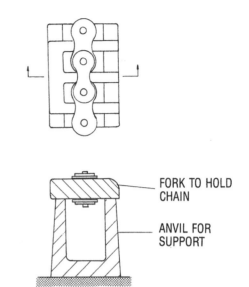

FORK TO HOLD CHAIN

ANVIL FOR SUPPORT

Fig. 7-37 A chain detaching tool.

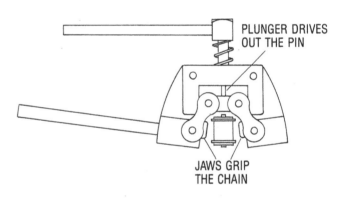

PLUNGER DRIVES OUT THE PIN

JAWS GRIP THE CHAIN

Fig. 7-38 A chain detacher that will drive out the link pin.

removed and if the atmosphere is dusty, regular cleaning becomes even more important.

The procedure recommended is as follows:
1 Remove the chain.
2 Check the chain and the sprockets for wear and corrosion.
3 Wash the chain in kerosene or similar cleaning fluid. Soaking may be necessary for a very dirty chain.
4 Drain off the cleaning fluid and soak the chain in lubricating oil.
5 Hang the chain and allow excess lubricant to run off.
6 Clean the sprockets and check alignment.
7 Reinstall the chain.

Lubrication

Proper lubrication is critical to the operation of chain drives and the achievement of satisfactory service life. The general principles of lubrication apply and

the following points should be remembered.
- Lubrication should be regular and the frequency determined by the operating conditions.
- Whatever lubrication system is employed, the lubricant must penetrate the chain joints.
- Chains should be cleaned regularly so that lubricant can flow into the joints.
- The higher the chain speed the greater the supply of lubricant required.
- Some means of protecting the chain from dirt and other contaminants should be provided.

The most common methods of chain drive lubrication are:

Manual   using a brush or oil can
            suitable for simple drives
Drip feed   low horsepower and low speed drives
Automatic   using an oil bath or an oil spray
            suitable for moderate to high speed
            drives

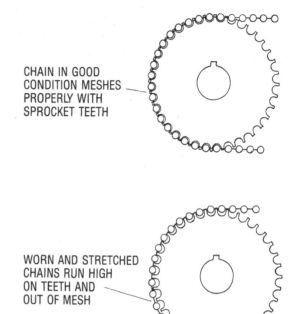

CHAIN IN GOOD CONDITION MESHES PROPERLY WITH SPROCKET TEETH

WORN AND STRETCHED CHAINS RUN HIGH ON TEETH AND OUT OF MESH

Fig. 7-39 Wear or excessive slack can cause the chain to ride high on the sprocket teeth.

FAILURE PATTERNS

The common symptoms and causes of chain drive malfunction and failure can be summarised as follows:

Operating symptoms

Noise

Chain drives are generally relatively noisy, certainly more so than belt drives, but if noise becomes excessive it may be an indication of malfunction. As for other indicators of the condition of machine elements, it is useful to establish an operating level when the system is new and properly adjusted so that changes in noise level can be detected. Intermittent ticking or slapping sounds may be associated with interference and should be investigated immediately.

Chain jumping

Chains can have a tendency to climb the sprockets and can reach the point where they jump off. This may be due to wear or to excessive slack, both of which allow the chain to ride high on the sprocket teeth. (Fig. 7-39)

Alternatively, excessive material build-up on the sprocket teeth may interfere with the correct mating of the chain and teeth.

Chain whipping

Too much slack, or pulsating loads, may cause the chain to whip. This may also occur if some of the chain links have become stiff or have seized.

Chain breakage

It is rare for a chain to break completely although this may occur if a condition of malfunction has existed for some time and excessive fatigue has taken place. It can also occur when a chain is badly worn. The lengthened chain may jump a tooth on the sprocket which causes excessive tension and consequent failure of the side plates.

Overheating

Excessive operating temperatures are an indication that the drive is operating at too high a speed, with too great a load, or with inadequate lubrication. It is often difficult to determine the temperature of a chain during operation but it can be checked as soon as the machine is shut down. Once again, a comparison with normal operating conditions must be made in order to detect evidence of malfunction.

Fig. 7-40 Evidence of chain wear.

Symptoms found on inspection

Once the drive is shut down and inspected the following symptoms may be evident.

Wear

Chains may show evidence of wear on the pins, bushings and rollers.

The amount of wear that can be tolerated will depend on how much elongation of the chain it causes. As a general rule, single pitch chains can tolerate up to around 3% elongation whereas double pitch chains can only tolerate up to 1½%. The manufacturer should be consulted for precise tolerances. If wear becomes excessive the chain may jump a tooth on the slack side of the sprocket resulting in excessive loads in the side plates.

If wear occurs on the insides of the side plates and also on the sides of the sprocket teeth then misalignment of the drive is indicated.

Wear to the sprocket teeth will often be evident in the form of 'hooked' teeth as shown in Fig. 7-41.

HOOKED TEETH
DUE TO WEAR

Fig. 7-41 Evidence of wear to the sprocket teeth.

Galling

If lubrication is inadequate at high loads or speeds the bearing surfaces of the pins and bushings may weld and then tear apart. The resultant effect is known as galling.

Fig. 7-42 Galling in a chain drive.

Corrosion

The presence of rust or surface pitting and roughening is evidence that corrosion is taking place.

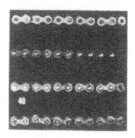

Fig. 7-43 Evidence of corrosion.

Link seizure

The links of a chain should be free to move without significant resistance so that the chain can mesh properly with the sprocket teeth. If the chain joints become tight or seize altogether this will affect the operation of the drive and may cause whipping or jumping of the chain. Tightness may be due to the build-up of foreign material or products of corrosion which prevent the lubricant from being effective. Damage to the side plates from interference or careless handling may also cause stiffness.

Causes of failure

The common causes of the symptoms listed above are as follows:

Misalignment

Correct alignment is extremely important. Misalignment of sprockets leads to uneven wear of chain and sprockets and also results in noisy operation.

Incorrect slack adjustment

If the slack in the chain is either too much or too little, the operation will be noisy. The chain may climb up on the sprocket teeth and tend to jump when the chain is too loose. Excessive wear can also be expected.

Lack of lubrication

Inadequate lubrication will affect the operation of chain drives just as it will other machine elements. Excessive noise and rapid wear will be the most obvious symptoms, with chain breakage being the ultimate result if no action is taken.

Material build-up

The build-up of foreign material on chain and sprockets due to the presence of dust or other contaminants will prevent the lubricant from being effective, and may interfere with the correct meshing of the chain and sprockets. Proper protection and regular cleaning will help to eliminate this problem.

Interference

Any interference between the chain and guards or other machine elements will cause wear and damage to the chain. It should be recognisable by the increase in noise level.

Vibration

Excessive vibration, transmitted along either shaft to the drive sprockets, will tend to increase wear and may cause failure of the chain fasteners.

Overload

Rapid wear, chain breakage and overheating may all occur if the drive is subjected to loads or speeds above the recommended design limit.

Worn Sprockets

Excessive wear or other damage to sprockets will hasten the deterioration of the condition of the chain. New chains should never be installed on worn sprockets.

Summary of the symptoms and causes of chain drive failure

Symptoms		Causes
Operating	Inspection	
Noise	Wear	Misalignment
Jumping	Galling	Slack adjustment
Whipping	Corrosion	Lack of lubrication
Breakage	Link seizure	Material build-up
Overheating		Interference
		Vibration
		Overload
		Worn sprockets

7.3 GEAR DRIVES

Gear drives are used to transmit power from one machine to another where changes of speed, torque, direction of rotation or shaft orientation are required. They may consist of one or more sets of gears depending on the requirements. In most cases the gears are mounted on shafts supported by an enclosed casing which also contains a lubricant.

Most gear drives in use are speed reducers. They reduce the speed of shaft rotation between driver and driven machines and, at the same time, produce a corresponding increase in torque. The recent increased use of high speed machinery, such as centrifugal compressors, has also generated a need for speed increasers.

PRINCIPLES OF OPERATION

A gear is a form of wheel with teeth machined around the outer edge which allow it to engage with another similar wheel or rack.

The most important features of a gear are the tooth profile or cross-sectional shape, and the number of teeth. In modern gears the tooth profile is based on an involute curve which is the shape produced when a line is traced by a point on a cord which is 'unwound' from a cylinder as shown in Fig. 7-44.

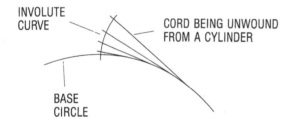

Fig. 7-44 Generating the involute form.

The advantage of the involute form is that when two gear teeth mesh together a constant velocity ratio is maintained with the minimum of sliding and the maximum of rolling action of one tooth against the other. This feature helps to reduce wear and extend the life of the gear.

In order to understand the geometry of gears it is useful to imagine that the simplest form of gears is two plain discs which touch tangentially. If sufficient friction exists between the surfaces in contact then there is no need for special teeth to be cut. However, there is a limit to the torque that can be transmitted by friction and so teeth are cut into the outer edges of the discs to provide a means of positive engagement as shown in Fig. 7-45.

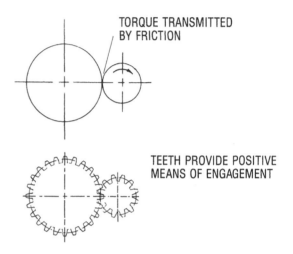

Fig. 7-45 Teeth provide a means of positive engagement.

The imaginary circles on which the gears are cut are called the pitch circles, and the pitch circle diameter is the major dimension on which gear geometry is based. The other important dimension is the pressure angle. This is the angle between a tangent to the pitch circle and the line of contact of two mating teeth as shown in Fig. 7-46.

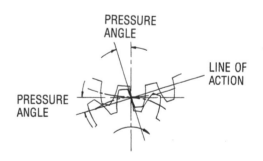

Fig. 7-46 The pressure angle.

A full explanation of the terms used to describe the geometry of a circular gear is given in Fig. 7-47.

If two gears are to mesh properly they must have the same pressure angle. Standard pressure angles of 14.5° and 20° are used with 20° being the most common.

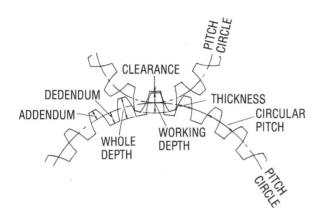

Fig. 7-47 Terms used in circular gear geometry.

In practice, gears are cut to provide running clearance between mating teeth. This is known as backlash.

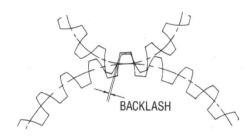

Fig. 7-48 Gears are cut to provide backlash.

The characteristics of mating gears are often described by the term, diametral pitch. This term refers to the ratio of the number of teeth to the pitch circle diameter of the gear and reflects the size and shape of the teeth. Hence two mating gears must also have the same diametral pitch as well as the same pressure angle.

There are several ways in which diametral pitch can be calculated.

- diametral pitch $= \dfrac{\text{number of teeth}}{\text{pitch circle diameter}}$

- diametral pitch $= \dfrac{\pi (3.142)}{\text{circular pitch}}$

- diametral pitch $= \dfrac{\text{number of teeth} + 2}{\text{outside diameter}}$

The speed relationship between two mating gears depends on the number of teeth on each gear and can be determined as follows:

$$\begin{matrix} \text{speed of driven gear} \\ \text{(RPM)} \end{matrix} = \begin{matrix} \text{speed of driver} \\ \text{(RPM)} \end{matrix} \times \dfrac{\text{no. of teeth on driver}}{\text{no. of teeth on driven}}$$

TYPES AND ARRANGEMENTS

The following types of gears are in common use.

Spur gears

The spur gear is the simplest type of gear and has teeth cut parallel to the axis. Spur gears may be used as external or internal gears or as a rack and pinion.

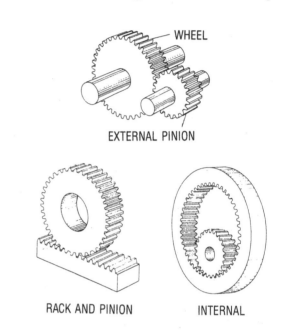

Fig. 7-49 Spur gears.

Spur gears are used to transmit power between parallel shafts operating at moderate speeds. They are simple to manufacture, do not develop end thrust and are the preferred type to be used where practicable.

It is conventional to refer to the large gear as the wheel or bull gear and the smaller as the pinion.

Helical and herringbone gears

Helical gears are also used to transmit power between parallel shafts but have the teeth cut on an angle.

Fig. 7-50 Helical gears have teeth cut on an angle.

The advantage of this design is that several teeth are in mesh at the same time and this results in greater load carrying capacity and smoother operation. Because of the angle of the teeth, helical gears produce end thrust which must be carried by the shaft bearings. This can be overcome by the use of two rows of opposed helical teeth in a 'herringbone' arrangement shown in Fig. 7-51. Herringbone gears are generally not recommended when externally applied end thrust is present or when operating speeds are very high because of the tendency for one helix to carry most of the load.

Fig. 7-51 Herringbone gears.

Bevel gears

Bevel gears are used to transmit power between two intersecting shafts, normally at right angles. The teeth on bevel gears may be plain or spiral.

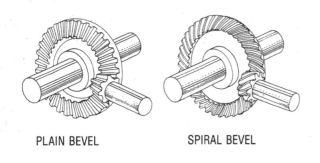

PLAIN BEVEL   SPIRAL BEVEL

Fig. 7-52 Bevel gear teeth may be plain or spiral.

Spiral bevel gears distribute the load over several teeth, in the same way that helical gears do for parallel shafts, and hence give smoother operation.

Hypoid gears

A variation of the spiral bevel gear is the hypoid gear which is designed to transmit power between two non-intersecting and non-parallel shafts.

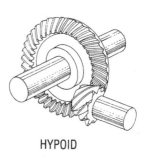

HYPOID

Fig. 7-53 A hypoid gear.

Worm gears

Worm gears are designed to transmit power between two non-intersecting shafts at right angles, as shown in Fig. 7-54, and are used when high ratio speed reduction is required. The worm may be cut with one or more threads and must be of the same hand as the wheel.

WORM

Fig. 7-54 Worm gear.

Whatever type of gear is employed, the arrangement may involve one or more pairs of gears depending on the degree of speed reduction required. (Fig. 7-55)

Most gear drives are mounted in fully enclosed casings but large ring gears may be installed as open gears with a suitable guard arrangement.

Gears are generally made from steel or cast iron and are surface hardened in order to increase the wear resistance.

MAINTENANCE PRACTICES

As with other methods of power transmission, the key to satisfactory operation of gear drives is good alignment, proper lubrication and the exclusion of dirt and other contaminants.

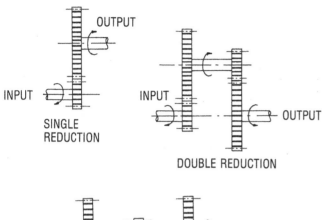

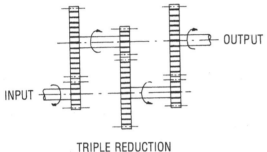

Fig. 7-55 The number of pairs of gears depends on the degree of speed reduction required.

Alignment

In most gear drives the alignment is determined by the machining of the casing and the bearing housings and under normal conditions the gears should be automatically aligned. Care should be taken when mounting a gear box to ensure that no distortion of

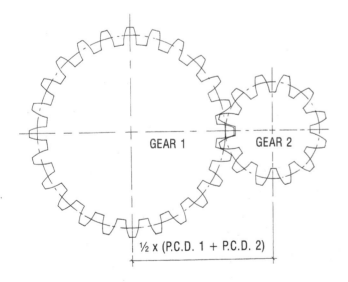

Fig. 7-56 Centre distance determines correct meshing.

the casing occurs when the mounting bolts are pulled down. (See Chapter 3.) Excessive wear and run-out in the bearings will result in run-out and wear of gear teeth and should be rectified as soon as possible.

When gears are open, or installed in such a manner that adjustment of the relative positions is possible, either the centre distance between the gears or the backlash between the teeth can be measured. The centre distance between two gears in proper mesh should be equal to half the sum of the pitch circle diameters of the gears. (Fig. 7-56)

If it is more convenient to measure the backlash this can be done using feeler gauges. Care must be taken, whenever the relative position of gears is adjusted, to make sure that the shafts remain parallel so that the gears run true. Parallel alignment of gear tooth faces can be checked by applying marking blue to the pinion teeth and then turning the gear wheel over by hand. The contact pattern on the gear wheel teeth should be even across the face.

Lubrication

Most gear boxes contain a reservoir of lubricating oil in which the lower halves of the gears are submerged. As the gears turn they pick up lubricant which protects the teeth during contact. If maximum gear life is to be achieved the correct lubricant must be used and the correct operating level must be maintained in the sump or reservoir.

A gear box should be checked from time to time for leaks and these should be corrected as soon as possible.

The products of gear tooth wear will collect in the oil reservoir along with any other contaminants which enter the gear box casing. It is therefore necessary to change the lubricant at regular periods as recommended by the manufacturer. This is particularly important during the run-in phase of the machine when the rate of wear-debris production tends to be relatively high.

Gear boxes that are pressure lubricated are normally fitted with a filter and this must be changed or cleaned periodically.

Open gears are usually lubricated by a sump in which the bull gear runs. If the atmosphere is dusty then a large build-up may develop on the gears and it must be cleaned off from time to time.

FAILURE PATTERNS

Like all machine elements which involve relative motion between lubricated components, there are a number of ways in which gear failures occur.

Operating symptoms

The symptoms of gear malfunction found during operation are relatively few in number and easily detectable.

Noise

Even when gears are in good condition they produce a significant amount of noise. This is because of the continuous impact of the gear teeth as they mesh with each other and it will vary with speed and torque transmitted. Every gear combination and gear box has its own distinctive running sound when it is operating satisfactorily and familiarity with that sound will assist the maintenance technician in detecting deterioration in the running condition. Gears in good condition should produce a constant hum with a relatively smooth tone. Once the gears begin to deteriorate, or some malfunction in their operation develops, this noise will change. The sound may become much rougher which could indicate that the gears are not properly in mesh or that they are out of alignment. Misalignment may also cause a rhythmic or pulsating sound to develop. When the surface condition of gear teeth begins to deteriorate, due either to wear or other factors, the sound of the gears tends to increase in pitch and develops into a whine.

Whatever the nature of the sound, whether it be a growl or a whine, the more it increases the more it is an indication of malfunction of the drive assembly. It is not possible to be precise about the nature of the sound that may emanate from a particular problem and the above description can only be considered as a general guide. Increase in noise level, however, can always be treated as positive evidence of a change in the condition of a gear drive.

Vibration

The nature of gears makes it inevitable that their operation is accompanied by a certain amount of vibration. As with noise levels, this will vary according to the type of gears and the speed and load transmitted. An increase in vibration levels will occur when condition deteriorates or a malfunction develops. The most likely causes of an increase in vibration levels are shaft misalignment and teeth running out of mesh. This may occur because of faulty assembly or deterioration in the condition of the shaft support bearings. Wear and deterioration of tooth surface condition, unless they become excessive, are less likely to cause an increase in vibration levels.

Overheating

Generally speaking, overheating of a gear drive is likely to be due to overload or inadequate lubrication and is more likely to occur with enclosed gear boxes than open gears. Serious misalignment or running out of mesh may also cause an eventual increase in running temperature. These conditions should be detected from the change in noise and vibration levels before this becomes significant.

Symptoms found on inspection

Once the external evidence of gear drive malfunction becomes significant then the unit must be shut down and examined so that further evidence can be gathered and cause of failure determined. The following patterns of failure are those most commonly encountered.

Wear

There are various ways in which gear tooth surfaces can wear. In the early stages of gear life, surface irregularities often cause pitting along the pitch line which later disappears when the gears wear in.

Mobil

Fig. 7-57 Pitch line pitting.

Caterpillar

Fig. 7-58 Normal wear under light to moderate loads.

Normal wear occurs because of metal-to-metal contact between mating teeth and under light to moderate loads will appear as shown in Fig. 7-58.

At higher loads and speeds the adhesion and welding of surfaces that takes place, due to the failure of the lubricant, becomes more extreme and the wear pattern that develops is known as scoring.

The most extreme form of this failure mode is known as galling.

When foreign particles such as dirt and grit are present abrasive wear patterns, similar in appearance to scoring, may develop.

The wear patterns that develop due to misalignment and running out of mesh will have quite specific characteristics regardless of the nature of the type of wear. Misalignment will cause an uneven wear pattern to occur across the tooth face as shown in the example in Fig. 7-61.

Running out of mesh will cause either undercutting of tooth faces due to interference (Fig. 7-62) or a shift in the location of the wear pattern towards the tips of the teeth.

Caterpillar

Fig. 7-59 Scoring.

Mobil

Fig. 7-61 Wear pattern from misalignment.

Caterpillar

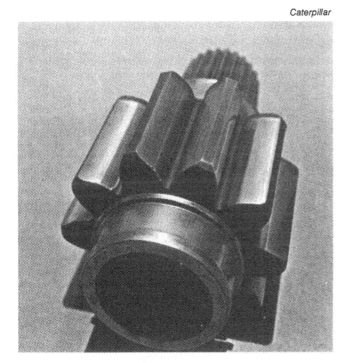

Fig. 7-60 Abrasive wear patterns.

Fig. 7-62 Undercutting of gear teeth.

It is also possible for corrosive wear to occur due to chemical attack from either contaminated lubricant or an additive. This will be evident by etching, not only of gear tooth surfaces, as shown in Fig. 7-63, but also of other gear surfaces.

Caterpillar

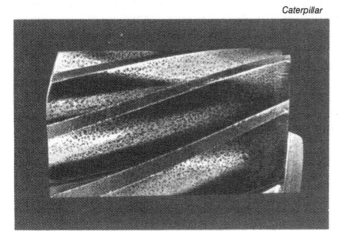

Fig. 7-63 Etching of gear tooth surfaces due to corrosive wear.

Fatigue

The mechanism of fatigue failure in gears follows the same patterns as for other machine elements. Cyclical stressing due to the intermittent contact of gear teeth, especially at high loads, causes sub-surface cracking which then develops into pitting and spalling as shown in Fig. 7-64.

Mobil

Fig. 7-64 Pitting and spalling caused by cyclical stress.

Plastic flow

Deformation of gear tooth surfaces due to the effect of heavy loading is referred to as plastic flow or cold flow. This is more likely to occur with soft, ductile gears but can also occur with case-hardened gears. The surface metal of the gear teeth flows under load and the effect is often described as rippling, ridging, rolling or peening.

Mobil

Fig. 7-65 Examples of plastic flow due to peening of tooth edges.

Tooth breakage

If regular inspection is carried out then complete or partial breakage of gear teeth should rarely occur. When it does happen there will usually have been some weakening of the teeth due to a condition that has been present for some time. Once a gear breaks it usually becomes inoperable. The two principal causes are fatigue and heavy impact loads. Fatigue failure is often associated with the presence of a

Mobil

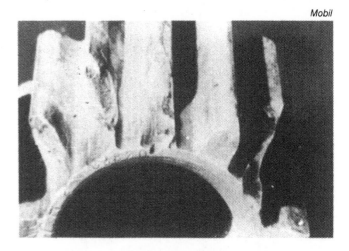

Fig. 7-66 Example of gear tooth breakage.

stress raiser such as a quenching crack or some surface defect such as a notch or a tool mark.

Causes

Inadequate lubrication

The operation of gears is vitally dependent on an adequate supply of the correct lubricant and if this is available gears should give almost unlimited life. If the wrong lubricant is used or if lubricant is allowed to deteriorate then the gears will begin to wear. It is also important that lubricant replacement is carried out according to the manufacturer's recommendation. Additives should be carefully selected to suit the particular operating conditions, and lubricant supply systems, including filters, should be properly maintained.

Misalignment

There are two different ways in which gears may be misaligned. They may be **out of parallel** or **out of plane**.

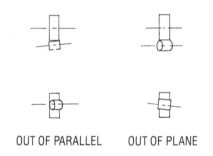

OUT OF PARALLEL   OUT OF PLANE

Fig. 7-67 Misalignment of gears.

The uneven wear patterns referred to in the previous section are produced by misalignment and if the condition is allowed to continue then tooth breakage may eventually occur.

Out of mesh

If the gears are set up so that the pitch circles are not touching each other then they can be considered to be out of mesh.

If the centre distance between the gears is too small and there is insufficient backlash then interference occurs and the tip of one tooth tends to dig into the root of the mating tooth and produce excessive wear. This will cause rapid deterioration of gear condition and may result in excessive noise and vibration. If gears are set too far apart the backlash will be excessive and wear will occur close to the tips

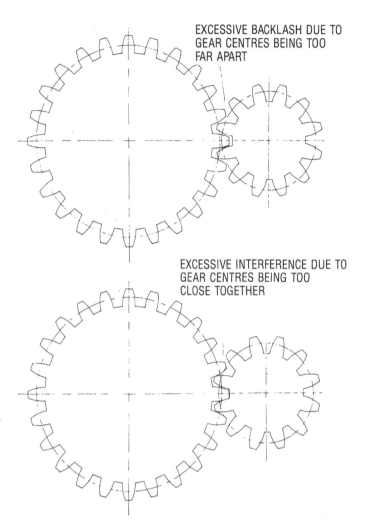

EXCESSIVE BACKLASH DUE TO GEAR CENTRES BEING TOO FAR APART

EXCESSIVE INTERFERENCE DUE TO GEAR CENTRES BEING TOO CLOSE TOGETHER

Fig. 7-68 Gears out of mesh.

of the teeth which may even break from the impact. Noise and vibration will also increase due to the large amount of backlash.

Overload

Speeds and loads which exceed design limits and high impact loads will accelerate wear processes and lead to the likelihood of premature failure. Heavy spalling or galling and tooth breakages are the usual consequences of overload conditions. The design limits of the drive should be checked and the unit operated according to the manufacturer's instructions.

Contamination

The contamination of gear lubricant due to the presence of dirt, dust or other abrasives will increase wear rates and cause scratching of tooth surfaces. Particles of wear metal or chips of broken teeth will also cause significant damage. Careful attention should be paid to the condition of seals and filters and the regular replacement of lubricant if contamination is to be avoided.

Moisture

The presence of moisture in a gear box may cause rusting to develop. In order to avoid moisture due to condensation build-up, special breather arrangements may be required whereas ingress of moisture from other sources should be prevented by the oil seals.

Lubricant breakdown

If lubricants are not replaced regularly they may deteriorate to the point where harmful acids may form. If corrosion is detected then lubricant analysis may be required to establish whether any change in properties of the lubricant has occurred. The effect of lubricant additives on particular metals should also be considered and the lubricant manufacturer consulted for advice.

Summary of the common symptoms and causes of gear drive failure

Symptoms		Causes
Operating	Inspection	
Noise	Wear: adhesive	Inadequate lubrication
Vibration	scoring	Misalignment
Overheating	abrasive	Out of mesh
	uneven	Overload
	corrosive	Contamination
	Fatigue/spalling	Moisture
	Plastic flow	Lubricant breakdown
	Tooth breakage	

7.4 SHAFT COUPLINGS

Couplings are the devices used to connect two shafts with a common axis of rotation so that one can drive the other. Unlike the other transmission elements discussed in this chapter, couplings do not change the characteristics of the motion they transmit. Speed, torque and direction of rotation all remain the same from driver to driven.

PRINCIPLES OF OPERATION

There are two major categories of shaft coupling. When two shafts are truly aligned the coupling may be **rigid** or **solid**, in which case one shaft merely becomes a direct extension of the other. If some misalignment of the shafts is likely, the coupling must contain a mechanism able to absorb that misalignment and still transmit motion smoothly from one shaft to another. Couplings which accomplish this are known as **flexible** couplings and they operate according to two basic principles.

The simplest types of flexible coupling employ a flexible element to absorb the misalignment. Elastomeric materials such as rubber are normally used for the flexible element although metal elements are also common. This type of coupling requires no lubrication and is generally referred to as **material-flexing**.

The second group may be described as **mechanical-flexing** and involves mechanical components that slide or otherwise move relative to each other to provide the necessary flexibility. This type of coupling requires lubrication to minimise wear.

TYPES

There are various commonly-used couplings which meet the needs of the majority of industrial applications.

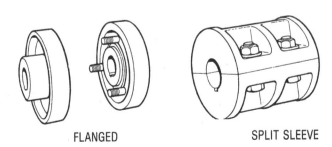

FLANGED   SPLIT SLEEVE

Fig. 7-69 Two common types of rigid coupling.   *Renold*

Rigid

The two most common rigid couplings are the flanged and sleeved types shown in Fig. 7-69.

Rigid couplings cannot normally be used where shafts need to be aligned on installation, as in the case of independent driver and driven units, but are used for line shafting where some means of disconnection is required.

Material-flexing

The most common types of material-flexing couplings are:

Jaw couplings

These utilise a flexible element which fits between two sets of metal jaws.

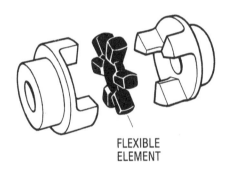

FLEXIBLE
ELEMENT

Renold

Fig. 7-70 Jaw coupling.

Pin couplings

Two metal flanges are connected by steel studs upon which flexible buffers, usually made of rubber, are mounted. They are sometimes referred to as crown pin or cone ring couplings and a typical example is shown in Fig. 7-71.

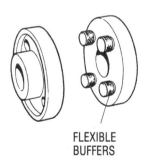

FLEXIBLE
BUFFERS

Renold

Fig. 7-71 Typical example of a pin coupling.

Disc couplings

In this type of coupling the flanged half bodies are connected with a flexible disc which is usually made of an oil resistant composite material.

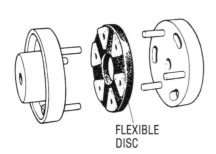

FLEXIBLE
DISC

Renold

Fig. 7-72 A simple disc coupling.

A more sophisticated type of disc coupling employs a laminated metal disc ring as the flexible element. In order to accommodate both angular and parallel misalignment, two disc elements are normally incorporated within the coupling.

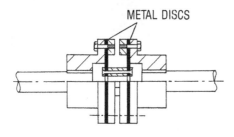

METAL DISCS

Fig. 7-73 Disc coupling with two disc elements.

Elastomeric couplings

Couplings that rely for flexibility on the elastomeric properties of a specially designed element are available in various designs. The most common is probably the type shown in Fig. 7-74.

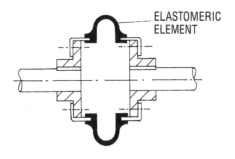

ELASTOMERIC
ELEMENT

Fig. 7-74 Common type of elastomeric coupling.

Mechanical-flexing

The most common types of mechanical-flexing couplings are as follows:

Gear couplings

The coupling hubs are machined with external gear teeth which mesh with an internal set of teeth machined into the flanged cover as shown in Fig. 7-75. Relative movement between the mating sets of gear teeth provides the necessary flexibility. Lubrication is essential and the covers are fitted with seals to contain the lubricant which is usually grease.

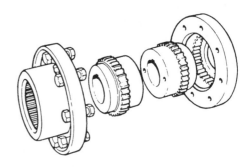

Renold

Fig. 7-75 A gear coupling.

Chain couplings

The coupling hubs are machined with sprocket teeth and the two halves are connected by a length of duplex roller chain.

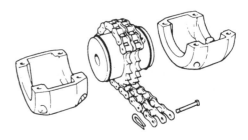

Renold

Fig. 7-76 Chain coupling.

An oil-tight cover is provided to contain the grease and protect the coupling.

Grid couplings

The hubs of grid couplings are machined with axial grooves or slots into which is fitted a flexible metallic grid.

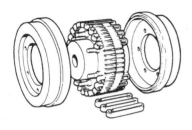

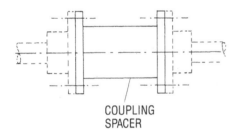

Renold

Fig. 7-77 A grid coupling.

A sealed cover contains grease to lubricate the coupling in a similar way to a chain coupling.

Spacer couplings

These are not a further type of coupling but represent a particular variation of the standard designs that enables length to be added to the coupling as shown in the example in Fig. 7-78.

COUPLING
SPACER

Fig. 7-78 Use of a spacer coupling to add length.

The principal advantage of a spacer coupling is that it allows the coupling hubs to be removed from the shafts without disturbing the position of either machine. Most large machines that cannot be easily moved are fitted with spacer couplings.

MAINTENANCE PRACTICES

The following general considerations are relevant to the maintenance of couplings.
- The most important factor in ensuring that satisfactory coupling performance is achieved is the alignment of the shafts. Although flexible couplings are designed to cope with misalignment there is always a limit to the amount that can be handled without causing rapid wear of the coupling.
- Before a coupling can be installed it must be dissassembled and it is wise to record the order of components so that it can be reassembled in the same order.

- It is vital that the coupling hubs are mounted securely on the shafts so that no relative movement between the hub and shaft can occur. The most common methods of securing the hubs are to use a key and keyway, an interference fit or a tapered bush arrangement.
- Coupling hubs are often supplied with only a pilot hole and must be machined out to fit the shaft. It is vital that the coupling is bored true and concentric otherwise the coupling may be thrown out of balance.
- Before fitting the coupling hubs, the shaft and the bore of the hubs should be inspected to ensure that they are free of nicks, burrs and other damage.
- A rotating coupling can be a safety hazard and should never be operated without a suitable safety guard. The guard should be robust and designed so that personnel cannot come into contact with the coupling during operation.

Lubrication

The lubrication of mechanical-flexing couplings is important if rapid wear is to be avoided. Coupling covers are provided with grease points and the manufacturer's lubrication recommendations should be followed. If machines operate in a dirty or dusty environment it is advisable to strip couplings down from time to time and clean them with a degreasing fluid. New grease should then be applied before restarting the machine.

Assembly

The major consideration when assembling a coupling is to ensure that the two halves are securely fixed to the shafts and that the necessary clearance gap exists between mating faces. Because the procedure varies for different types of couplings it is wise to consult the manufacturer's recommendations. The use of tapered bushes of the type shown in Fig. 7-79 allows

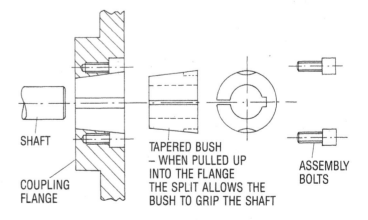

SHAFT

COUPLING
FLANGE

TAPERED BUSH
– WHEN PULLED UP
INTO THE FLANGE
THE SPLIT ALLOWS THE
BUSH TO GRIP THE SHAFT

ASSEMBLY
BOLTS

Fig. 7-79 A tapered bush arrangement.

a standard machined coupling to be fitted to a range of shaft sizes by selection of the appropriate bushing.

Alignment

The factor that most significantly affects the performance of shaft couplings is shaft alignment. Because of its importance this topic has been treated separately and is the subject of Chapter 8.

FAILURE PATTERNS

A properly assembled, aligned and lubricated coupling should give virtually unlimited service. Failures occur from time to time, however, and the maintenance technician should be aware of the likely symptoms and common causes.

Operating symptoms

The symptoms of coupling malfunction are relatively few and obvious.

Noise

Under normal operation a coupling should be noise free. If coupling hubs are not tight and wear begins to develop then the coupling may develop a clicking or a rattling sound.

Vibration

Once a coupling begins to vibrate, not only its condition but the condition of other machine elements, such as bearings and seals, will begin to deteriorate. Loose hubs, excessive wear and misalignment are all potential causes of coupling vibration. It is also possible that vibration may be transmitted from other parts of the machine and this too can cause excessive wear or other damage to a coupling.

Symptoms found on inspection

If the performance of a coupling indicates a malfunction then the machine should be shut down at the earliest opportunity and inspected. The following symptoms are common evidence of malfunction. Because there are many different types of couplings the treatment given here is only general.

Wear

The nature of wear shown by a coupling will depend on its construction but excessive wear is usually evidence that the coupling has not been operating satisfactorily. Material-flexing couplings will show evidence of wear to the flexible element and if this is

excessive there may even be damage to other components such as pins and flanges. Mechanical-flexing couplings will show evidence of wear to the mechanical elements which accommodate the relative movement. Hence the teeth of gear couplings must be inspected as must the chain and sprocket teeth of chain couplings. Wear of grid couplings is likely to show up in the axial slots that hold the grid element. Wear may also be observed in the bore of the coupling, and to the shaft, if the coupling hub has not been securely fixed.

External damage

The coupling should run well clear of any other machine parts and external damage may indicate interference with the guard or other machine elements. This may appear as extreme wear or scoring on the outside of the coupling.

Material damage

The flexible elements of material-flexing couplings are often made of rubber and may be susceptible to attack by oil and other liquids. Swelling and loss of elasticity will be the likely result.

Causes

The common causes of coupling failure can be summarised as follows.

Misalignment

As mentioned above if misalignment is too great then the coupling will wear rapidly.

Inadequate lubrication

If mechanical-flexing couplings are not adequately lubricated the life of the coupling will be significantly reduced.

Improper assembly

Failure to mount the coupling hubs securely on the shafts and failure to properly install covers and seals will lead to early malfunction.

Interference

Coupling guards should be designed to give adequate clearance so that there is no danger of interference.

Contamination

Excessive build-up of dirt around the coupling may cause it to bind and prevent if from moving to accommodate whatever misalignment may be present.

Overload

If the operating conditions exceed the design limits of the coupling it is most likely that rapid wear and premature failure will result.

Summary of the failure patterns associated with shaft couplings

Symptoms		Causes
Operating	*Inspection*	
Noise	Wear	Misalignment
Vibration	External damage	Inadequate lubrication
	Material damage	Improper assembly
		Interference
		Contamination
		Overload

7.5 CLUTCHES

Clutches are devices that enable two shafts or rotating elements to be connected or disconnected while at rest or in relative motion. They must be capable of transmitting the maximum torque requirement of the drive system, and of disengaging completely and allowing one shaft to rotate independently of the other. Some clutches allow transmission of motion in one direction only.

The best known application of a clutch is in the transmission system of a motor vehicle. There are many other industrial applications in which clutches are critical machine elements, and they play a particularly important role in the operation of automatically controlled machinery.

PRINCIPLES OF OPERATION

There are two aspects to clutch operation: torque transmission and clutch actuation.

Torque transmission

There are three principle ways in which torque and motion are transmitted from one shaft to another: positive engagement, friction and wedging action.

Positive engagement

The simplest and most basic arrangement is one which relies on positive engagement by means of teeth on coupling halves such as the one shown in Fig. 7-80. Engagement is achieved by allowing one of the clutch halves to slide axially along the shaft. This mechanism has the advantage of being unable to slip but is very limited in its ability to allow engagement on the run, although this can be achieved at low speeds with certain tooth designs.

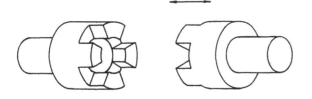

Fig. 7-80 Engagement by teeth.

Friction

Many types of clutch rely on friction to transmit torque. This is the principle used in most motor vehicle clutches. A typical arrangement involves bringing two contacting surfaces together and relying on the friction between them to transmit the torque. Often a special plate covered with material with a high coefficient of friction is interposed between the surfaces to increase the efficiency of the clutch.

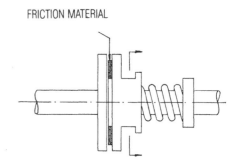

Fig. 7-81 Engagement by friction.

The torque carrying capacity of a friction clutch is directly related to the area of contact between the friction surfaces.

Wedging action

The principle employed in freewheeling and over-running clutches, where motion is required to be transmitted in one direction only, is one that relies on the wedging action of a roller or specially designed element trapped between two races. Rotation of the drive shaft in one direction causes the rollers to wedge and transmit torque to the other shaft, as shown in Fig. 7-82. If the drive shaft rotates in the opposite direction, the driven shaft can either remain stationary or freewheel.

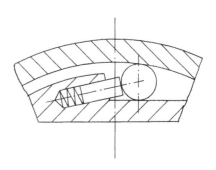

Fig. 7-82 Engagement by rollers.

A variation of this principle involves the use of specially designed wedges called sprags. These are held between concentric races by springs, as shown in Fig. 7-83.

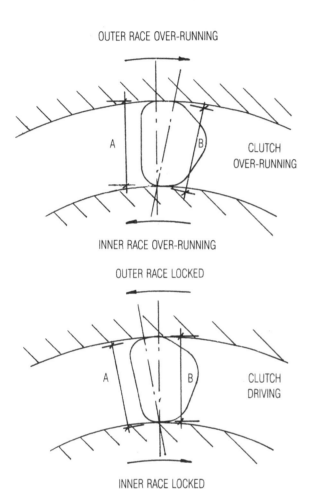

OUTER RACE OVER-RUNNING

A   B   CLUTCH OVER-RUNNING

INNER RACE OVER-RUNNING

OUTER RACE LOCKED

A   B   CLUTCH DRIVING

INNER RACE LOCKED

Fig. 7-83 Engagement by sprags.

The sprags are formed with one set of dimensions, 'A', less than the radial gap between the races and the other, 'B', greater than the gap. Rotation of the inner shaft in a clockwise direction will cause the sprag to wedge and drive the outer shaft in the same direction. Similarly, rotation of the outer race in an anticlockwise direction will cause the inner race to rotate. If the inner shaft rotates in an anticlockwise direction, however, the sprag will take up a position such that its shorter dimension will allow it to disengage and the outer race will remain stationary.

Clutches that utilise this principle also allow the driven shaft to 'over-run' the drive shaft by rotating faster and in the same direction.

Clutch actuation

There are four principal ways in which a clutch mechanism can be actuated. They are **mechanical**, **pneumatic**, **electrical** and **hydraulic**. The method used and the design of the system will depend on the application and the conditions under which the system is required to operate.

TYPES AND APPLICATIONS

There are two general industrial applications in which clutches are used: power transmission, and indexing, over-running and backstopping.

Power transmission

The following types of clutches are the ones most commonly used for power transmission.

Dog-tooth

The dog-tooth clutch is a positive displacement type that is very simple in design. One half is operated by a lever and slides on a key or splines. It can be operated only when stationary or moving at very low speeds.

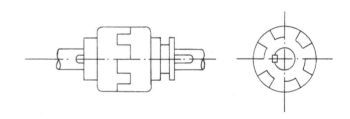

Fig. 7-84 Dog-tooth clutch.

Single-plate friction

This employs a plate with friction material on both sides which is clamped between two steel plates when the clutch is engaged. The friction plate is mounted on splines or a key on the output shaft and can slide axially. The pressure that causes the drive to engage is usually provided by a set of springs, and disengagement is achieved by a lever that acts against the springs and releases the pressure plate.

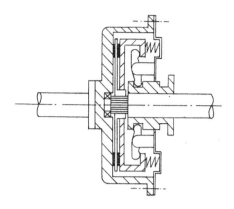

Fig. 7-85 Single-plate friction clutch.

Multi-plate friction

This is a development of the single-plate clutch and has several friction plates (usually five, seven or nine) keyed to one shaft. Interleaved with them are steel pressure plates keyed to the other shaft. As with the single-plate clutch, the pressure is provided by a set of springs or sometimes a single spring.

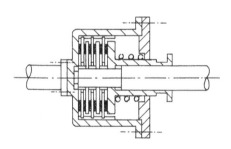

Fig. 7-86 Multi-plate friction clutch.

Cone

A cone clutch has two conical mating surfaces, one of which is usually lined with high-friction material. The torque transmitting capacity of a cone clutch is greater than that of a flat plate clutch of the same diameter because of the increased area of contact.

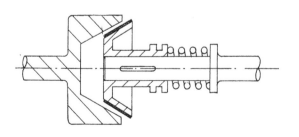

Fig. 7-87 Cone clutch.

Centrifugal

This type is used when engagement of a load has to be achieved at a particular rotational speed. A typical design has spring-loaded weights mounted in radial slots in a member connected to the drive shaft. As the speed of the drive shaft increases, the weights, which are faced with friction material, are thrown out against the surface of a drum mounted on the output shaft. This is the type of clutch commonly used on chain saws.

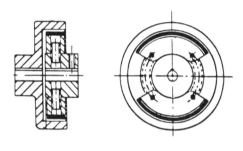

Fig. 7-88 Centrifugal clutch.

Dry fluid

This is a type of centrifugal clutch in which metal particles, such as steel shot, are compacted under the action of the centrifugal force produced by rotation. The particles are contained in a hollow member in which a disc attached to the driven member rotates.

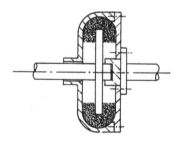

Fig. 7-89 Dry fluid clutch.

Fluid coupling

In a fluid coupling clutch both input and output shafts carry impellers which have radial vanes. The vane spaces are filled with oil which circulates in the vanes when the coupling rotates. The input wheel acts as a pump and the output wheel as a turbine so that power is transmitted. There is always a loss of speed in this type of coupling due to slippage.

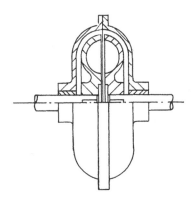

Fig. 7-90 Fluid coupling.

Indexing, over-running and backstopping

There are many industrial applications where a clutch is required to fulfil functions other than, or in addition to, power transmission.

Indexing is the transmission of torque on an intermittent basis either from a continuously rotating shaft or by converting the reciprocating linear motion of a pneumatic or hydraulic cylinder into rotary motion. This function is typically used for indexing conveyors, packaging materials and sheet steel in presses where precise location is required.

Over-running allows the output to rotate faster than the input and is typically used in starter drives for engines or turbines to provide automatic decoupling of starter motor drive when operational speed is reached.

Backstopping is the prevention of undesirable reverse rotation. This is achieved by anchoring one race of the clutch so that the other race is free to rotate in one direction but is immediately locked in reverse. Backstopping is typically used on conveyors, elevators, cranes and pumps.

There are two principal types of clutch that are used in such applications: roller and sprag.

Roller

The freewheeling roller clutch permits a shaft to be driven in one direction only and operates according to the principles previously described.

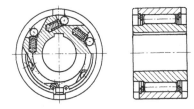

Fig. 7-91 Roller freewheeling clutch.

Sprag

This type, also described above, utilises the wedging action of sprags in an annular space.

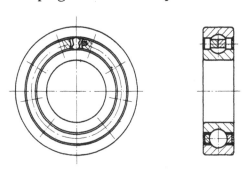

Fig. 7-92 Sprag freewheeling clutch.

Contact free sprag

A variation on the sprag clutch, the contact free sprag is designed such that the sprags disengage totally when the clutch is freewheeling or over-running. The sprags are pivoted so that the action of centrifugal force lifts them clear of the stationary or slower running race. They can be designed to be either internally or externally disengaging.

MAINTENANCE PRACTICES

The maintenance practices associated with clutches are dependent to some extent on the type of clutch concerned. There are a number of general guidelines that should be followed.

Alignment

As with other drive components, the alignment of the two shafts is critical to the operation of a clutch. In the case of coupling type clutches, the requirements are comparable with those for couplings and will normally be specified by the manufacturer. Plate clutches, such as those fitted in vehicles, are also sensitive to misalignment and run out of the flywheel, and bearing housings may need to be checked against manufacturer's recommendations.

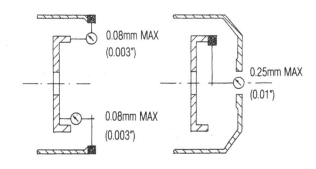

Fig. 7-93 Checking run-out for plate clutches.

Assembly

Assembly should not be attempted unless alignment has already been checked.

The assembly of coupling type clutches requires that the halves be firmly secured to the shafts, usually with parallel keys. Grub screws are often used to prevent sideways movement.

When assembling plate clutches, care should be taken to ensure that no grease or oil comes into contact with the linings. The splines or keyway on which the clutch plate is mounted should be given a light coating of lubricant.

Roller and sprag clutches may be assembled in a number of different ways depending on the application. The major considerations are alignment, secure location, accurate positioning and the avoidance of any undue axial loading on clutch bearings.

Lubrication

Roller and sprag clutches require correct lubrication in order to achieve maximum service life. Generally they will be oil lubricated although grease is also used under special conditions. Under no circumstances should lubricants containing molybdenum disulphide or graphite or similar additives be used as these reduce the coefficient of friction and prevent locking of the clutch at the point of torque transmission.

The engagement of tooth clutches can be assisted with a light coating of lubricant. Other types of clutches do not normally require lubrication except where they contain bearings and these will usually be sealed.

FAILURE PATTERNS

Clutches are required to perform a demanding role and carry fluctuating loads under difficult conditions. They are a potential source of maintenance problems if not properly installed and serviced.

Operating symptoms

Because of the important function it performs, any malfunction of a clutch will usually be very obvious and affect the performance of the drive system.

Noise

When a friction clutch is in good operating condition, it should not generate noise. There may be a slight change in tone between engagement and disengagement due to the loading of a thrust bearing.

Roller and sprag clutches should be silent when driving but may generate noise when freewheeling or over-running. As with other machine elements, it is change in noise that indicates a possible malfunction. The noise generated by a clutch under normal operation and in good condition should be established when it is first put into service. Any significant deviation may then be treated as a sign of malfunction.

Shudder

A clutch may shudder or chatter due to grabbing of the friction surfaces. This may cause vibration of the whole transmission. There are a number of conditions that may cause this problem. With friction plate clutches the condition of the friction material and the pressure plate surfaces may deteriorate due to wear or distortion. Oil on the friction material may also glaze and interfere with operation. Loose coupling halves, misalignment and wear of rollers and sprags may also cause a drive to shudder.

Slipping

Most clutches will slip briefly at the point of engagement but should then run without slipping. A slipping clutch may result in either a loss of speed of the driven unit, an over speeding of the drive, or both.

Overheating

A properly engaged clutch should not generate heat. Overheating may be associated with slipping or bearing failure.

Non-engagement

It is possible that a clutch may not engage at all. This would suggest a problem with the actuating system or a jamming of the clutch mechanism.

Non-release

Non-release of a clutch may be a result of problems with the actuating system. It may also be due to binding of the driven plate or half on shaft splines or keyway. An accumulation of dirt or corrosion in roller and sprag clutches may cause a freewheeling mechanism to bind.

Symptoms found on inspection

Once a clutch begins to malfunction it should be taken out of service to avoid further damage. The following are the symptoms most likely to be found on inspection.

Wear

The most common cause of plate clutch malfunction is wear of the friction material that is used to achieve engagement of the drive. Under normal circumstances, when the clutch has been operating correctly, wear should be uniform across the linings. Uneven wear may suggest misalignment or distortion of the pressure plate. The pressure surfaces will also exhibit signs of wear after extended operation and this will be accelerated if the linings are not renewed at the appropriate time. Wear may also be evident on the face of the thrust bearing.

Wear of keys or splines may cause misalignment and may be a contributing factor to uneven wear of clutch linings.

Careful inspection may be required to reveal evidence of wear of rollers and sprags in freewheeling clutches. The annular races of sprag clutches and the wedging faces of roller clutches may also wear under normal operation.

Oil on linings

Evidence of oil on the friction material linings of a clutch would explain slipping. If the oil has burnt to a glaze then this may explain shudder or grabbing of the clutch.

Mechanical damage

External damage to a clutch may be evidence that it has been struck by some object. The effect of this on alignment should be checked.

Damage to clutch components during assembly or as a result of improper operation may lead to operating problems and further damage.

Friction plate clutches should be carefully inspected for damage to pressure plate surfaces, pressure springs, thrust race and contact surface, splines, keyways and shaft surfaces, and the operating mechanism.

Freewheeling clutches should be inspected for damage to rollers and sprags and the contact races. The mechanism should also be inspected for broken or damaged springs.

When a clutch contains bearings they should also be inspected for damage and for evidence of wear.

Causes

The likely causes of clutch failure are as follows.

Misalignment

As mentioned above, misalignment of shafts will result in uneven wear and premature failure.

Inadequate lubrication

Freewheeling clutches will wear rapidly and fail prematurely if the correct lubrication is not provided.

Improper assembly

When assembling plate clutches it is important that pressure springs are properly located and that housing spigots are properly aligned.

Failure to mount clutch hubs properly on keyed shafts or splines will cause malfunction due to misalignment and vibration.

Interference

Clutches that rotate with the drive should not be allowed to contact a guard or other machine elements.

Contamination

Leakage of oil into a friction plate clutch assembly from shaft bearings will seriously affect operation. Oil seals should be maintained in good condition to protect the clutch mechanism.

Dirt build-up in a clutch mechanism will prevent free movement of components and interfere with operation.

Overload

The application of excessive torque to a clutch is likely to cause rapid wear and lead to premature failure. In the case of some clutches this may be preceded by slipping or other malfunction.

Summary of the failure patterns associated with clutches		
Symptoms		Causes
Operating	Inspection	
Noise	Wear	Misalignment
Shudder	Oil	Inadequate lubrication
Slippage	Mechanical damage	Improper assembly
Overheating		Interference
Non-engagement		Contamination
Non-release		Overload

SHAFT ALIGNMENT

Perfect alignment is seldom achieved but good alignment can and should be possible for any machine of which continuous, reliable operation is demanded. Poor alignment can cause vibration and wear and lead to premature failure of bearings, seals, couplings and other machine elements. Despite the fact that flexible couplings can tolerate quite significant degrees of misalignment, it is recommended that shaft alignment should always be as accurate as possible as any misalignment present will set up vibration patterns that will be transmitted through the machine and tend to cause wear. Alignment is a critical aspect of machinery operation and the techniques used should be properly understood.

PRINCIPLES OF ALIGNMENT

Misalignment

Shafts can be misaligned in two ways.

Parallel misalignment

This occurs when the shaft centre-lines remain parallel, but are offset, as shown in Fig. 8-1.

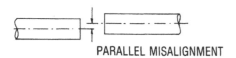

PARALLEL MISALIGNMENT

Fig. 8-1 Parallel misalignment.

Angular misalignment

This occurs when the shaft centre-lines are out of parallel, although they may intersect at the coupling, as shown in Fig. 8-2.

ANGULAR MISALIGNMENT

Fig. 8-2 Angular misalignment.

It is usual to find that both types of misalignment occur simultaneously and the alignment techniques described are designed to overcome both.

Tolerances

Depending on the size and type of the coupling, coupling manufacturers may indicate that parallel run-out of up to 0.5mm (0.020″) and angular run-out to 15° may be tolerated. It is important to realise that these figures represent the coupling tolerance and not the machine tolerance. Machinery manufacturers normally supply a recommended tolerance for a particular machine and some companies have their own standards which may be even more stringent. The higher the speed and the power transmitted, the more critical alignment becomes.

Adjustment

It is normal practice to adjust the position of the driving machine relative to the driven machine, especially when the driving machine is an electric motor with no interconnecting pipework. The position of driven machines is normally determined by associated process piping. It will have been set up to eliminate pipe strain and should not be disturbed.

Ideally the driving machine should be free to move in any direction with respect to the driven machine. However, if both are mounted directly onto the baseplate without shims, then it is clearly impossible to move the driving machine down with respect to the driven machine. Therefore before the alignment procedure is started a preliminary shimpack of, say, 1.25mm (0.050″) should be installed under both machines. These shims should be installed before the interconnecting pipework is aligned and bolted up.

Alignment in the vertical and horizontal directions should be carried out separately to avoid confusion and error. It is normal practice for vertical alignment to be carried out first otherwise the horizontal alignment may be disturbed when vertical adjustment is made.

Dial indicators

The operation of indicators should be clearly understood before they are used in shaft alignment. The important points to remember are that when the plunger is depressed, the gauge reads positive and when it is extended the gauge reads negative.

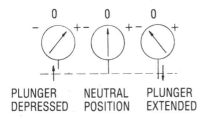

Fig. 8-3 Reading a dial indicator.

It is also important to remember that when a dial indicator is zeroed prior to taking readings, the plunger should be approximately in the mid-range position so that it is free to move in either direction.

Total indicator run-out

When a reading is taken with a dial indicator over 180°, i.e., from top to bottom of a shaft or from side to side, the difference in readings is known as the total indicator run-out (TIR).

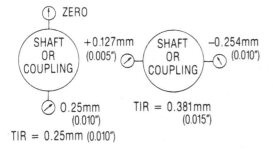

Fig. 8-4 Total indicator run-out.

For these readings to be properly interpreted it must be understood that total indicator run-out represents twice the shaft centre-line offset. As the indicator is turned through 180°, the shaft offset is doubled on the indicator scale, as shown below:

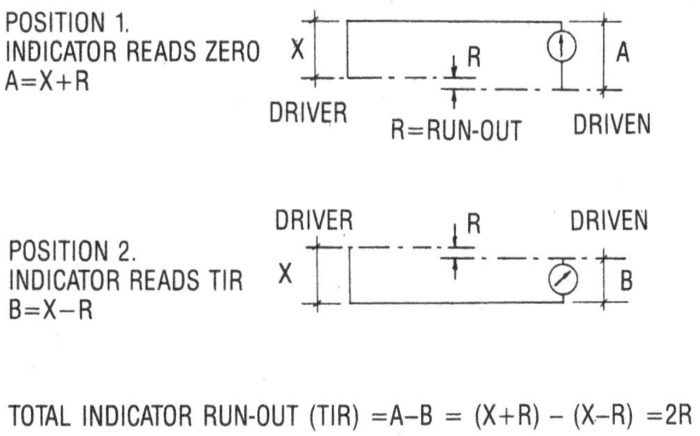

POSITION 1.
INDICATOR READS ZERO
$A = X + R$

POSITION 2.
INDICATOR READS TIR
$B = X - R$

TOTAL INDICATOR RUN-OUT (TIR) $= A - B = (X + R) - (X - R) = 2R$

Fig. 8-5 Total indicator run-out = 2 x shaft run-out.

Thermal growth

Because of the difference in thermal expansion between the driving and the driven machine during operation, shafts often need to be set up cold with an offset that disappears when the machines are up to operating temperature. This occurs most commonly with steam turbine drivers. Because a turbine undergoes considerable thermal expansion during operation, the shaft needs to be set lower than that of the driven machine in the cold state.

Information about thermal growth is normally supplied by the manufacturer but otherwise can be measured by setting up dial indicators on a fixed reference and recording the actual growth at the operating temperature.

It is generally accepted that a 'hot-check' for machinery alignment is of little value. Not only is it costly and time consuming to bring a machine up to temperature, stop it and determine alignment before it cools down, but the results are highly questionable.

Dowelling

In order to ensure that shaft alignment is maintained when the machine is in operation it is advisable that the position of both machines is secured by dowelling. Two dowels should be used for each unit and they should be located as far apart as possible. The dowel holes should be drilled and reamed in position through the machine mounting into the bedplate. The dowel pins should be parallel ground to give a light drive fit in the holes. The size of the dowels will depend on the size of the machine.

METHODS OF ALIGNMENT

The straight edge and feeler gauge method

The simplest and easiest method, but also the least accurate, is to use a straight edge across the coupling halves to test for parallel run-out and feeler gauges or a taper gauge between the coupling halves to test for angular run-out as shown in Fig. 8-6.

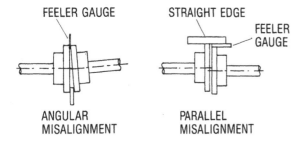

Fig. 8-6 Using a straight edge and feeler gauge.

Adjustments to the shimpacks under the driving machine will need to be made on the basis of trial and error, although repeated alignment work on an individual machine will help the technician to develop a feel for the adjustment required.

It is recommended that, with this method, angular run-out be eliminated first and then the remaining parallel run-out can be eliminated by making equal adjustments at both mountings of the driving machine.

This method is only recommended for use on machines fitted with flexible couplings capable of tolerating up to 15° angular run out, and 0.25mm (0.010″) parallel run-out, and for low horsepower, slow speed conditions. The degree of accuracy achieved is unlikely to be consistently better than 0.125mm (0.005″).

The face and rim method

This method can only be used for couplings installed with a spacer, and involves the use of two dial indicators supported on a bracket or two brackets set up across the coupling, and mounted on the driven machine as shown in Fig. 8-7. Note that the coupling spacer has to be removed for this method to be used.

As shown in Fig 8-7, parallel run-out is measured by readings taken on the outside diameter of the coupling and angular run-out is measured by readings taken on the face of the coupling half. The support brackets on which the dial gauges are mounted must be rigid and securely attached to the coupling halves. The alignment procedure is carried

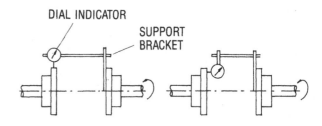

Fig. 8-7 The face and rim method.

out by first zeroing both indicators at top dead centre and then turning the shaft supporting the gauges through 360° and recording the gauge readings at 90° intervals. A typical set of readings would appear as follows.

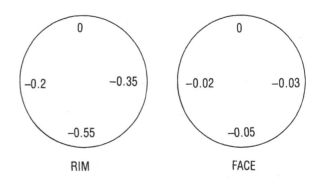

Fig. 8-8 A typical set of readings.

The readings taken on the coupling o.d. or rim represent the parallel run-out in both the vertical and horizontal directions. In this case the driving machine has been recorded as 0.275mm (0.011″) high at the coupling (remember shaft centre-line offset is half the total indicator run-out), while the run-out in the horizontal plane is 0.075mm (0.003″). The readings taken on the face of the coupling show the angular run-out which, in this case, can be seen to be greater in the vertical plane than the horizontal.

Although it is theoretically possible to use the readings obtained to calculate the precise adjustment required at the mountings of the driving machine, this is rarely, if ever done in practice. The procedure for adjustment, as with the straight edge and feeler gauge method, is essentially one of trial and error, backed up by the experience of the technician. Consequently several sets of readings may be required with progressive adjustments being made until satisfactory alignment is achieved.

The face and rim method has traditionally been one of the most widely used alignment techniques and is the specific method recommended in many maintenance manuals for rotating machinery. However, for high speed, high horsepower machinery where alignment tolerances become more critical, this method has a number of shortcomings.

Shortcomings of the face and rim method

- The indicator readings obtained by this method reflect not only the misalignment of the shafts but also the out-of-roundness, eccentricity and surface imperfections of the coupling upon which the readings were taken. Even with good quality, high speed couplings, run-out in excess of 0.025 mm (0.001″) in the coupling o.d. and face is not uncommon. These errors can be eliminated by determining the geometry of the coupling hub or by turning both shafts together but these precautions are difficult and rarely taken.
- To obtain accurate face measurements it is vital that end float is accounted for. The only way to do this is to ensure that both shafts are hard up against the thrust bearing at all times. Consistent readings are difficult to obtain however, because the axial position of the shaft will depend on how hard it is pushed.
- Face readings are taken across a relatively small diameter and the angular misalignment measured here can be greatly magnified at the outboard end of a long shaft. For example a difference of 0.05 mm (0.002″) measured across a 100mm (4″) coupling, which might normally be considered as acceptable may represent an error of as much as 0.75mm (0.030″) at the rear end of the machine shaft.
- Removal of the coupling spacer is time consuming and increases the risk of coupling damage.

These problems make the face and rim method unsuitable for high speed machinery and have also resulted in it being superseded in general applications by the reverse indicator method which has gained increasing popularity in recent years.

The reverse indicator method

This method involves using dial gauges to take readings on the outside diameter of the coupling hubs (or the shafts) only. No face readings are involved. The gauges are supported on separate brackets mounted on opposite halves of the coupling so that a reading is obtained on each. (Fig. 8-9)

The main advantage of this method is that it allows the relative position of the two shafts to be accurately determined. It is then possible to determine the exact adjustment required at the mountings of the driving

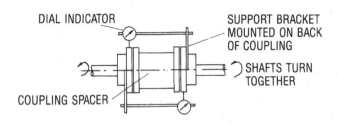

Fig. 8-9 Typical arrangement of the dial gauges.

machine to produce proper alignment. The shortcomings of the face and rim method are overcome in the following manner.

- By turning both shafts simultaneously, errors caused by coupling hub run-out are entirely eliminated.
- Since no face readings are involved, the end float of shafts is of no concern.
- By spanning the entire coupling, angular misalignment is greatly magnified and can be measured with greater accuracy.
- If the support brackets are designed to be mounted on the back of the coupling halves, the coupling spacer does not need to be removed.

One problem that affects the reverse indicator method as well as the face and rim method is the problem of support bracket deflection. This is a matter of concern and should not be ignored if accurate cold alignment is to be achieved. The problem can be minimized by ensuring that the brackets are as rigid as possible, and, if necessary, the deflections can be measured and appropriate corrections made. Deflection of the bracket can be measured by mounting it on a piece of barstock held in a lathe chuck as shown in Fig. 8-10. With the bracket at top dead centre of the barstock the gauge is set to zero. The assembly is then rotated through 180° and the gauge is read at the bottom position.

The difference in indicator readings reflects the deflection in the bracket. The figure can be used to correct readings taken during machine alignment.

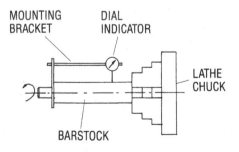

Fig. 8-10 Measuring deflection of the bracket.

Because of the superior features of the reverse indicator method, the procedure is described in detail on a step-by-step basis.

1 Set up the dial indicators on suitable support brackets, preferably so that both indicators read on the same side of the coupling. This is not essential, but saves time in taking readings.

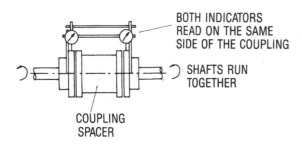

Fig. 8-11 The recommended arrangement for setting up the dial indicators.

2 Adjust both indicators to zero at top dead centre of the coupling hubs.
3 Check that the indicator plungers are in mid-range and readjust if necessary.
4 Check that the indicators can be rotated through 360° without interference.
5 Adjust the driving machine to a roughly central position in the horizontal plane with respect to the driven machine. Indicator readings, if zeroed at top dead centre, should be roughly equal at the 90° and 270° positions, as shown in Fig. 8-12.

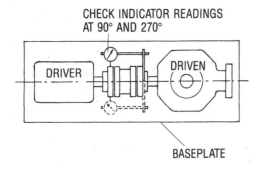

Fig. 8-12 Centralising the driver.

This step ensures that the indicator readings taken during the vertical alignment process are taken across the full diameter of the coupling or shaft. If the two machines are not roughly in line, then the vertical readings taken will give a false

value of total indicator run-out as shown in Fig. 8-13.

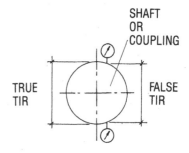

Fig. 8-13 Difference between false TIR and true TIR.

6 Measure the distance between the indicator plungers and the distance between the indicator on the driven machine and the centres of the mounting bolts on the driving machine as shown in Fig. 8-14.

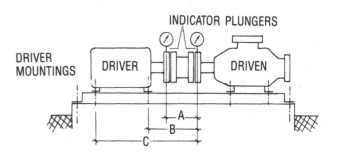

Fig. 8-14 Machine dimensions.

7 Plot the measurements A, B, and C on a sheet of graph paper as shown in Fig. 8-15. A suitable horizontal scale will need to be selected according to the size of the machine.

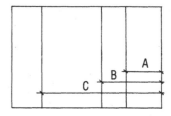

Fig. 8-15 Dimensions marked on graph.

The right hand vertical axis now represents the position of the indicator reading on the coupling hub of the driven machine. The other positions marked on the graph represent the position of the indicator readings on the coupling hub of the driving machine and the front and rear mounting positions of the driving machine.

8 Check the zero reading on the indicators at top dead centre, and then rotate both indicators through 180°. Record the total indicator run-out for each indicator at bottom dead centre.

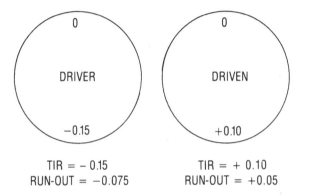

TIR = − 0.15   TIR = + 0.10
RUN-OUT = −0.075   RUN-OUT = +0.05

Fig. 8-16 Sample readings.

9 Determine the relative position of the driving machine shaft with respect to the driven machine shaft by halving the total indicator run-out and determining which shaft is high at each point of measurement. See the example given in Fig. 8-17.

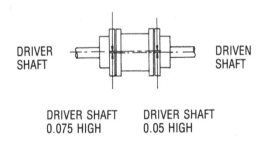

DRIVER SHAFT   DRIVER SHAFT
0.075 HIGH   0.05 HIGH

Fig. 8-17 Relative shaft positions at points measured on coupling.

The question of which shaft is high can be resolved by considering whether the indicator reading is positive or negative at bottom dead centre. The following chart shows the various possibilities.

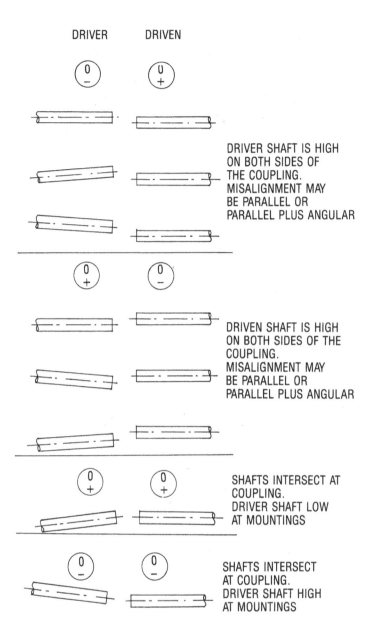

Fig. 8-18 Possible relative shaft positions.

10 Draw a horizontal line on the graph to represent the centre-line of the driven shaft.

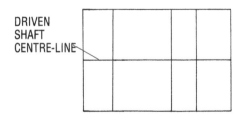

DRIVEN
SHAFT
CENTRE-LINE

Fig. 8-19 Driven shaft centre-line shown on graph.

11   Plot the positions of the driving machine shaft on the vertical axes of the graph corresponding to the point of measurement as shown in Fig. 8-20.

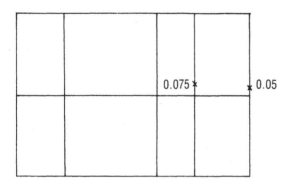

Fig. 8-20 Plotting the position of the driving shaft.

12   Join the two points together, and extrapolate the line back to the positions of the driving machine mountings. The line drawn now represents the position of the driving machine shaft centre-line relative to the driven machine shaft centre-line. The amount of adjustment at both front and rear mountings required to bring the shafts into line can now be scaled off the graph as shown. In the example given, the motor is too high at both front and rear mountings.

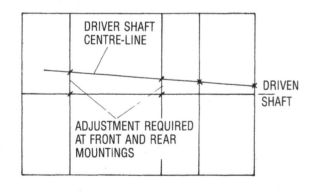

Fig. 8-21 Scaling off the required adjustment.

13   Adjust the shimpacks under the driving machine mountings accordingly, retighten the mounting bolts, and then recheck the alignment as before. Repeat the process until the required accuracy is achieved.

14   Repeat the whole procedure to obtain alignment in the horizontal plane. This time the indicators should be set to zero at one side of the coupling and total indicator run-out read at the other.

Jacking screws should be used to adjust the position of the driving machine, and it may be necessary to set up an additional dial indicator, using a magnetic mounting, to measure the movement of the machine, as shown in Fig. 8-22.

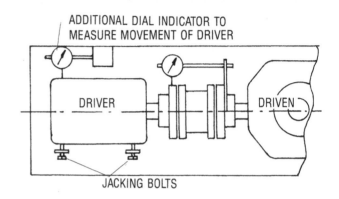

Fig. 8-22 Measuring movement of the machine with an extra dial indicator.

15   When both horizontal and vertical adjustments are complete, a final set of readings at 0°, 90°, 180° and 270° should be taken, with the indicators zeroed at 0°, as a record of the relative shaft positions. Readings should be recorded in the following manner, as if facing the respective coupling halves.

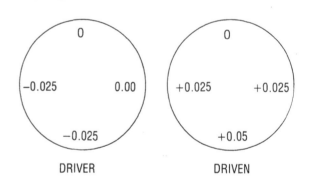

Fig. 8-23 Method of recording readings.

General hints on alignment

- Check the tools. Make sure that dial indicators do not stick.
- Use good quality shimpack material. If the material is damaged, discard and replace with new material.
- Find a good way to turn the shafts. For larger machines, make up a clamp-on fixture if necessary.

- If a mechanical end face seal is fitted to the machine, disengage the faces before turning the machine shafts.
- Beware of environmental changes that may affect alignment. Machines exposed to direct sunlight may undergo non-uniform thermal growth.
- Make sure that indicators are positioned to read squarely on the shaft.

- Always check the readings taken. If the readings cannot be repeated, they are not acceptable.
- Stop the shaft at precise 90° increments and turn the shaft in one direction only. If you pass the 90° mark, go round again.

SEALS

Seals are used to prevent the leakage of liquids, solids and gases from items of rotating machinery and other types of industrial equipment and to stop dirt and other sources of external contamination from entering a machine or piping system.

As industrial equipment has become more sophisticated the conditions under which seals have to operate have become more arduous and sealing technology has developed rapidly in the last twenty years in order to meet these developing needs. The demands of the aerospace industry have had a particularly strong influence on the advancement of seal technology.

Despite the sophisticated developments that have taken place, traditional seal designs are still found in the vast majority of industrial applications.

Before examining particular types of seals it is useful to consider how seals can be classified according to their operating characteristics and application.

The most fundamental distinction to be understood is the difference between static seals and dynamic seals. **Static seals** are those used for sealing surfaces between which there is no relative motion whereas **dynamic seals** are used where relative motion occurs. Static seals are often referred to as **joints** whereas dynamic seals may be referred to as **packing** or **gland packing**.

Dynamic seals can be further grouped into two main categories, **contact seals** where the sealing element rubs against a mating surface under load, and **clearance seals** which operate with positive clearance between the sealing surfaces.

The family tree of seals (Fig. 9-1) shows the common groups of seals found in rotating machines.

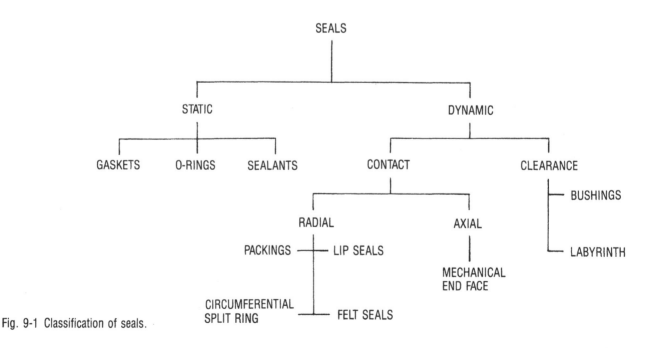

Fig. 9-1 Classification of seals.

9.1 GASKETS

A gasket is a static seal used to contain liquids, solids and gases in all types of machinery, containers and piping systems and may operate under a wide variety of service conditions. Gaskets are normally located between rigid and usually metallic sealing faces.

SEALING PRINCIPLES

A gasket requires an externally applied compressive force to maintain sealing contact. This force is normally provided by a set of flange bolts but various types of clamps or clips may also be used. The compressive force applied to the gasket material between the flange faces must be capable of:

- Accommodating surface variations in the flange faces and in the gasket material itself;
- Overcoming the hydrostatic end-force caused by the internal pressure trying to push the flanges apart;
- Leaving sufficient residual stress to contain the pressure and prevent it from extruding the gasket through the clearance space.

Fig. 9-2 shows the relationship between the forces acting on a gasket.

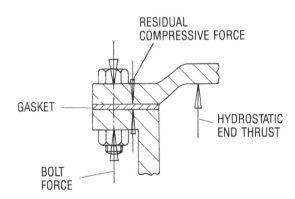

Fig. 9-2 Relationship between the forces acting on a gasket.

To ensure that the gasket continues to seal as the material relaxes in service it is recommended that the residual compressive stress be maintained well above the internal pressure to be contained. A factor of two is usually applied. Hence the compressive load on a gasket can be calculated from the relationship:

residual compressive stress = 2 x internal pressure

This then enables the bolt force or assembly load to be calculated from:

bolt force = hydrostatic end-thrust + residual gasket load

Because a safety factor of two has been allowed in determining gasket stress, no specific allowance for accommodating surface imperfections need be made. However, the ability of a gasket to accommodate surface imperfections will depend on the nature and thickness of the material in relation to the surface finish of the flanges and the internal pressure. A soft gasket material, which may provide a good seal against a comparatively rough surface, may extrude at the working pressure required. It may be necessary to improve the surface finish of the flanges so that a thinner, harder gasket material, capable of containing the working pressure, can be used.

Generally speaking, thinner gasket materials are more suitable for containing higher internal pressures when used between suitable machined flanges because the area exposed to the internal pressure is reduced, and thus the force tending to extrude the gasket is also reduced. See Fig. 9-3.

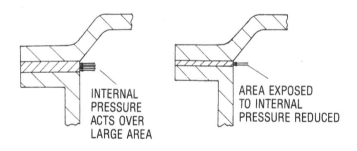

Fig. 9-3 Advantage of thinner gasket materials.

Although a high degree of surface finish would appear to be desirable in all cases, this can be unnecessarily expensive and it should be remembered that some surface roughness helps to provide grip, especially for harder gasket materials.

If good sealing is to be achieved, it is also important that flange faces are parallel and sufficiently rigid to resist distortion.

Although flange design can vary widely, there are a number of standard flange arrangements in common use in most industries and these are shown in Fig. 9-4.

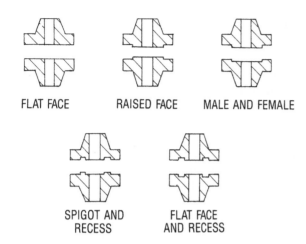

FLAT FACE   RAISED FACE   MALE AND FEMALE

SPIGOT AND RECESS   FLAT FACE AND RECESS

Fig. 9-4 Standard flange arrangements.

Gaskets may either be the ring type, where the outside diameter is located inside the bolt circle, or the full face type, as shown in Fig. 9-5.

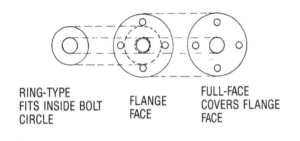

RING-TYPE FITS INSIDE BOLT CIRCLE   FLANGE FACE   FULL-FACE COVERS FLANGE FACE

Fig. 9-5 Full-face and ring type gaskets.

The spacing of flange bolts is related to the rigidity of the flange and should be designed to ensure even distribution of gasket load. In most cases, standard flanges are used which take these factors into account.

GASKET MATERIALS

A gasket material generally needs to have the characteristics of softness and deformability, and a high degree of elasticity and flexibility. The material should also be resistant to hardening and fatigue. The softness and deformability enable the gasket to accommodate the surface irregularities, and elasticity is necessary if the material is to be capable of responding to the slight change of shape required as loading fluctuates and flange bolts stretch.

Although the maintenance technician is not normally required to select gasket materials, some understanding of material properties and limitations is important.

Paper

A low cost material whose properties are improved when impregnated with fillers such as wax. Used most commonly in the automotive industry for sealing water, oil and petrol. They can be used up to 120°C (230°F) and 800 kPa (120 psi).

Cork

Useful in low load situations where flange faces are uneven. It has good resistance to oil and solvents but can be adversely affected by water. Can be used up to 50°C (125°F) and 350 kPa (50 psi).

Rubber bonded cork

The properties of cork can be improved by bonding it with an elastomer such as neoprene or nitrile. This produces a material with higher strength and flexibility that resists extrusion. It can be considered as a good general purpose material for low to moderate loads but is not recommended for highly alkaline or acidic conditions. It can be used from −30°C to 150°C (−22°F to 300°F) and up to 350 kPa (50 psi).

Rubber

This is one of the most versatile gasket materials especially when reinforced. The common property of elastomers is their ability to resume their original shape after deformation. For instance, rubber gaskets can be distorted, if required, during assembly. Although there are many types of rubber they all possess a number of common properties including compressibility and sensitivity to extreme temperature. Both high and low temperatures can seriously affect the properties of most types of rubber. The following types are commonly available:

Natural rubber Excellent mechanical properties but limited by low chemical resistance. Will also deteriorate in sunlight and ozone.
Buna-S A suitable substitute for natural rubber with greater resistance to water and heat.
Buna-N Has improved strength and abrasion resistance and also heat resistance. It is particularly suitable for use with mineral oils and some hydrocarbons but is soluble in others and cannot be used with organic acids.
Neoprene Has similar mechanical properties to natural rubber but has improved ozone resistance. Can be used with oils and non-aromatic hydrocarbons and can be made flame resistant.
Butyl Although its mechanical properties are not

as good as natural rubber it has high resistance to most chemicals and is widely used for gasketting. It is not recommended for petroleum-based fluids but has good high and low temperature properties.

Viton This is a fluorinated rubber that performs well with most chemicals except esters and ketones. It has good high temperature performance and ozone resistance.

Silicone rubber This is resistant to water and sunlight and can operate at both high and low temperatures. It is not suitable for use with hydrocarbons or at high pressures.

Compressed asbestos fibre (CAF)

This was the most common gasket material, until the health hazards of asbestos became known, because of its high temperature properties and resistance to chemical attack. It is made up of asbestos fibres bonded with an elastomer and has good mechanical properties. It can be used in most application other than with strong mineral acids.

Plastics

Many different synthetic materials have been developed in recent years and particularly since the limitations of asbestos became evident. One which is commonly used as a gasket material is PTFE (polytetrafluoroethylene or Teflon). PTFE is virtually chemically inert and can be used over a wide temperature range from −190°C to 250°C (−310°F to 480°F). The limitation of PTFE is that it has a tendency to cold flow and as a result is most commonly used in the form of an envelope that fits over a gasket cut from another material such as CAF or metal.

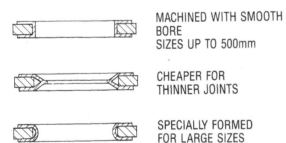

MACHINED WITH SMOOTH BORE
SIZES UP TO 500mm

CHEAPER FOR THINNER JOINTS

SPECIALLY FORMED FOR LARGE SIZES

Fig. 9-6 Typical configurations of PTFE envelope gaskets.

Various synthetic fibres, such as aramids, are now used in compounded form with synthetic rubber, such as nitrile, to replace the traditional CAF gasket material.

Metals

When operating conditions become too severe for the kinds of non-metallic gaskets mentioned above, various types of solid metal and semi-metallic gaskets can be used.

Solid metal gaskets made of materials such as lead, aluminium, copper, brass, monel, nickel and alloy steels are used when temperatures and pressures become extreme. They are usually used in a form that concentrates the load over a small area to increase the seating stress rather than as a flat ring.

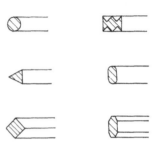

Fig. 9-7 Solid metal gasket sections.

Semi-metallic gaskets come in a number of forms the most common of which are metal clad and spiral wound.

Metal clad gaskets consist of a soft material, such as CAF or millboard, enclosed in a metal covering made from brass, aluminium, copper, monel or stainless steel. They can be made in intricate patterns and are commonly used as cylinder head gaskets and, in the process industry, for heat exchangers and pressure vessels.

Spiral wound gaskets are the most versatile of the semi-metallic type because of their outstanding recovery characteristics. They consist of a V-section metal strip wound into the form of a spiral with a non-metallic filler as shown in Fig. 9-8.

The metal strip is usually made of stainless steel but may be monel, nickel or titanium and the filler may be asbestos, PTFE or graphite. Retaining rings,

FILLER

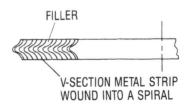

V-SECTION METAL STRIP WOUND INTO A SPIRAL

Fig. 9-8 Spiral wound gasket.

either internal or external, are used as shown in Fig. 9-11 to support the spiral wound section and to provide a stop against which the flanges may be tightened to give the gasket the correct amount of compression.

Spiral wound gaskets can operate at temperatures between −250°C and 1000°C (−420°F to 1830°F) and at pressures from vacuum up to 35 000 kPa (5000 psi). They are extremely versatile, and with careful selection of materials are suitable for use with the majority of fluids.

MAINTENANCE PRACTICES

There are a number of general factors that should be considered when removing or installing gaskets.

- It is advisable to have the new gasket cut or fabricated and ready to install before breaking a joint.
- Do not, if possible, make the gasket by hammering on the flange face. This can damage both the material and the flange.
- Use as thin a gasket as possible for the joint conditions.
- For full face gaskets the bolt holes should be the same size as the holes in the flange.
- The gasket inner diameter should be larger than the inside bore of the joint face to prevent the gasket interfering with the fluids contained. The amount will depend on the material. For example, rubber will swell more than CAF and hence greater clearance will be required.
- If a joint has to be broken regularly then a coating of graphite or similar dry lubricant on one or both surfaces may make the gasket easier to remove. If a lubricant is used then a check should be made to make sure it is compatible with the contents of the machine, vessel or pipeline.
- Gaskets on doors and lids that have to be opened frequently can be cemented on one side and smeared with lubricant on the other. The cement chosen must be able to stand up to the operating conditions.

Disassembling a joint

1 Before starting make sure the joint is isolated and that all valves are closed. Drain any residual liquid from the joint and purge any gas if necessary.
2 For piping flanges, loosen and remove all bolts if the gasket is full-faced. For ring type gaskets, loosen all bolts but only remove enough to remove the gasket.
3 In equipment flanges it is recommended that all bolts be removed.

4 Where necessary, spring flanges apart using flange spreaders. If wedges are used care must be taken not to damage the flanges.
5 When the gasket has been removed, clean the joint faces and remove all traces of the old gasket and any jointing compound used.
6 Examine the joint faces for any evidence of scratching, corrosion, erosion or distortion of any kind.

Installing the gasket and assembling a joint

1 Ensure that joint facings are clean and free from burrs.
2 Bolt or stud threads should be clean and lubricated and spot facings on the back of flanges should also be clean.
3 Insert enough bolts in one flange to locate the gasket and make sure it lines up evenly all the way around the inside.
4 With the gasket in place on one flange, bring up the mating flange. Every effort should be made to ensure the flanges remain parallel as they are brought together.
5 Vertical heavy flanges such as those on vessels and heat exchangers should be jockeyed into position using a crane or hoist. These should be positioned on four bolts at an angle of 90° to each other that can be pulled up evenly to allow the flange to find its seat.
6 Insert the remaining bolts and pull them up in the correct sequence as shown in Fig. 9-9.

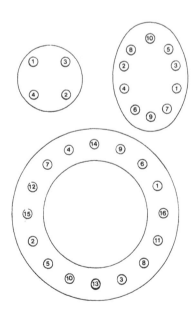

Fig. 9-9 Sequence for tightening bolts.

Do not snug up bolts on the first go round as this can tilt flanges out of parallel. If using an impact wrench, set for about half final torque on the first go round. Ensure that final tightening is uniform.

7 For the best performance in high temperature service make sure that bolts are retightened after 24 hours and then again after one week.

Special considerations for spiral wound gaskets

- Considerable care should be taken in handling these gaskets. Do not remove them from the cardboard backing until they are ready to be installed.
- Rough handling may break the spot welds that hold some of the rings together and the spiral wound section to the support ring.
- A spiral wound gasket should be sized so that the inner and outer wraps are in contact with the flange faces with clearance from the edges as shown below in Fig. 9-10.

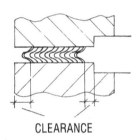

CLEARANCE

Fig. 9-10 Sizing of a spiral wound gasket.

- A spiral wound gasket gives a little as each bolt is tightened and does not have quite the same feel as other gaskets. Hence it is important that the joint is tightened in small steps to ensure the flanges do not tilt and damage the gasket.
- The flanges should be pulled down snug together for a recessed gasket or snug on to the retaining ring to give the correct gasket compression as shown in Fig. 9-11.

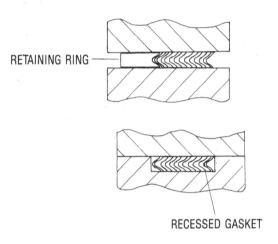

RETAINING RING

RECESSED GASKET

Fig. 9-11 Achieving the correct gasket compression.

9.2 O-RINGS

The elastomeric O-ring is one of the most versatile forms of static sealing arrangement and can also be used as a dynamic seal. The O-ring itself is normally contained within a groove machined into one of the flange faces. The elasticity of the material allows a good seal to be achieved with relatively low contact force.

SEALING PRINCIPLES

A key factor in the performance of an O-ring is the elastomeric properties of the material. An elastomeric material is one which can be repeatedly stretched to twice its normal length and still return to size on release. An O-ring with such properties, once compressed, produces an automatic tightening force and can also adjust to any deformation in the housing as long as the initial compression is maintained. An O-ring will cease to be effective when the material loses its elastomeric properties or when the initial compression of the gasket is lost.

Fig. 9-12 A typical arrangement of a static sealing O-ring.

As can be seen in Fig. 9-12 the O-ring is subjected to an initial compression which is controlled by the size of the ring and the depth of groove. Thus the initial sealing pressure is a result of this compression. When internal pressure is applied through the clearance gap between the flanges this deforms the O-ring further and forces it against the side of the groove, as shown in Fig. 9-13.

As the internal pressure is applied to the O-ring the elastomeric material responds in a manner which is known as self-energizing. This means that the

Fig. 9-13 The O-ring is forced against the side of the groove.

sealing pressure rises in direct response to the increase in internal pressure. However, the total sealing pressure is always higher than the internal pressure by an amount equal to the initial compression of the O-ring:

sealing pressure = internal pressure + initial compression

The maximum internal pressure that can be tolerated by an O-ring will depend on the characteristics of the material and its resistance to extrusion. If the internal pressure is too great then there may be a tendency for the O-ring to extrude into the clearance gap between the flanges as shown in Fig. 9-14.

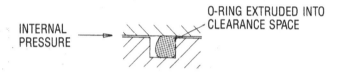

Fig. 9-14 O-ring extrusion.

This is most likely to occur with a static seal if pressure pulsations are sufficient to force the flanges apart. The tendency for an O-ring to extrude can be overcome by the use of a wedge shaped anti-extrusion ring, as shown in Fig. 9-15.

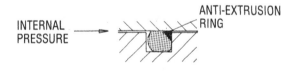

Fig. 9-15 Using an anti-extrusion ring.

MAINTENANCE PRACTICES

The most important concern when handling O-rings is to protect them from damage and to ensure that the correct loading is applied. The following general considerations should be taken into account.
- Ensure that the correct size of seal ring is used in relation to the size of groove.
- Make sure that grooves and recesses are clean and free from sharp edges and burrs.

- Care should be taken to ensure that the ring sits correctly in the groove and cannot get pinched between the flange faces.
- Flange faces should be pulled down evenly and to the correct pre-load recommended by the manufacturer.

FAILURE PATTERNS

The condition of an O-ring after disassembly may provide evidence of the cause of failure.

A ring that has been extruded will show the effects of nibbling along the i.d. as shown in Fig. 9-16.

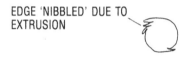

EDGE 'NIBBLED' DUE TO EXTRUSION

Fig. 9-16 Evidence of extrusion.

Damage caused to O-rings during assembly will usually be evident as nicks or cuts or possibly as twisting.

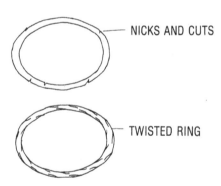

NICKS AND CUTS

TWISTED RING

Fig. 9-17 Evidence of damage during assembly.

An overheated O-ring will lose its elasticity and go hard and crack whereas a ring exposed to the wrong chemicals will swell and distort. See Fig. 9-18.

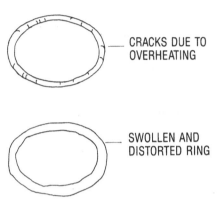

CRACKS DUE TO OVERHEATING

SWOLLEN AND DISTORTED RING

Fig. 9-18 O-ring deterioration.

9.3 SEALANTS

Sealants or liquid gaskets are specially formulated to contain liquids and gases under pressure and can be used as an alternative to gaskets and O-rings in operating conditions that are not too severe. They are relatively inexpensive and are available with a wide range of chemical resistance.

SEALING PRINCIPLES

A sealant performs quite differently from a gasket and allows metal-to-metal contact of the two opposing surfaces. Sealant materials are high viscosity liquids with adhesive properties that fill the voids caused by the surface asperities of the faces as shown in Fig. 9-19.

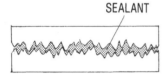

Fig. 9-19 Sealants fill the voids caused by surface asperities.

The fact that a sealant allows metal-to-metal contact to occur means that there is less likelihood of the joint working loose because of the relaxation of conventional gasket material. Once the joint is tight it should remain so and not require readjustment.

The use of a sealant can also reduce or eliminate the machining operations required to produce the degree of surface finish required for gaskets. Sealant materials have sufficient tackiness and viscosity to ensure they fill the voids between the surfaces if they are properly applied.

MATERIALS

The properties of sealant materials can be classified according to their hardening properties. Three basic types are available.

Non-hardening These are mastic type materials with limited adhesive properties. They contain plasticizers to make sure they remain soft.

Hardening-flexible There are various materials available, including neoprene, butyl, acrylic, silicone and polyurethane, that contain curing or setting adhesives but remain flexible when cured.

Hardening-rigid Epoxy, polyester and other resin compounds provide a seal that is rigid when cured.

The majority of sealants used fall into the category of hardening-flexible and common examples include the polyester urethanes used as general jointing compounds and silicone materials used as form-in-place gasket materials.

MAINTENANCE PRACTICES

Because sealants are invariably supplied as a proprietary brand the instructions for application are usually provided by the manufacturer. The important consideration from a maintenance point of view is to ensure that the correct type of sealant is used and that it is correctly applied.

There are several important points to remember in the application of sealants:

- Ensure that surfaces are prepared in the manner prescribed. Dirty or greasy surfaces usually interfere with the performance of the sealant.
- Ensure that the application of the sealant is even and, if recommended, applied to both joint faces. For other than pipe threads, make sure that joint faces are brought together square and that no dirt is allowed to enter the joint.
- Check the curing time of the sealant and make sure that no load is applied to the joint until the sealant has set.
- Do not use old materials that have exceeded their recommended shelf life.

9.4 MECHANICAL END FACE SEALS

Since their development as coolant system seals for the automotive industry, mechanical seals have now become one of the most widely used types of dynamic seal and have undergone considerable technological development. The range of conditions under which a mechanical seal can operate has been significantly increased by the use of new materials and innovative design features. Although relatively simple in construction, they involve complex design features and a high degree of precision in manufacture that necessitates high quality maintenance.

Mechanical seals have been specifically designed to prevent leakage of fluids from between rotating shafts and their housings and as such have replaced the use of soft packings in many situations.

PRINCIPLES OF OPERATION

The primary sealing function of a mechanical seal is achieved by two sealing rings with contacting faces, one of which rotates with the shaft and the other which is fixed in the housing. The contacting faces are lapped to provide an adequate seal and are held together by an axial force created by a mechanical device, such as a spring, and by the hydraulic force of the fluid contained.

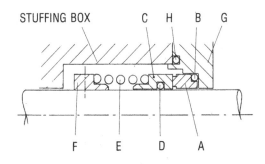

Fig. 9-20 A typical mechanical seal.

It should be noted that there are two secondary seals within a mechanical seal both of which are essentially static in nature and usually involve the use of O-rings. A seal has to be provided between the rotating face and the shaft and another between the stationary face and the seal plate as shown in Fig. 9-20.

The essential components of the mechanical seal are:

A Stationary seal ring This is usually made of carbon and is fitted into a machined recess in the seal plate.

B Stationary seal ring seal This is usually an O-ring which prevents leakage of fluid between the stationary seal ring and the seal plate.

C Rotating seal ring This is locked to the rotating shaft and forced against the stationary seal ring by the combined effort of the spring and the hydraulic pressure of the fluid. Common materials include stainless steel and tungsten carbide.

D Rotating seal ring seal This may also be an O-ring and prevents leakage between the rotating ring and the shaft. It also allows some freedom of movement of the rotating ring to ensure that full face contact of the seal faces is maintained. An alternative to the use of O-rings is the use of a PTFE wedge as shown in Fig. 9-21.

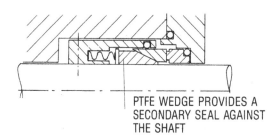

PTFE WEDGE PROVIDES A SECONDARY SEAL AGAINST THE SHAFT

Fig. 9-21 Using a PTFE wedge to prevent leakage between the rotating ring and the shaft.

E Spring A single coil spring is often provided as a means of loading the seal faces.

F Thrust collar This is attached to the shaft by a grub screw and provides a means of driving the spring and a stop against which the thrust is carried.

G Seal plate Comparable to the gland plate in a packed gland. This is the stuffing box cover plate which is also used to carry the stationary seal ring.

H Seal plate seal The seal plate gasket can also be considered as a secondary sealing point.

In addition to the physical characteristics of a seal there are a number of other features that should be understood.

● The contacting seal faces are lapped to precise limits to ensure that they are flat and are supplied in matching pairs.

● The primary sealing faces rely on a supply of lubricating fluid which may be the fluid contained by the seal or may be a secondary fluid supplied from an external source.

• The combination of mechanical and hydraulic loading on the seal faces can be varied by changing the hydraulic balance of the seal. In the seal arrangement shown in Fig. 9-20 the seal faces are subject to the full effect of the stuffing box pressure. This is shown more clearly in the simplified diagram Fig. 9-22.

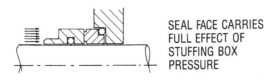

SEAL FACE CARRIES FULL EFFECT OF STUFFING BOX PRESSURE

Fig. 9-22 Effect of stuffing box pressure on seal face.

By changing the geometry of the rotating seal ring it is possible to achieve either partial or full balance of the hydraulic pressure as shown in Fig. 9-23.

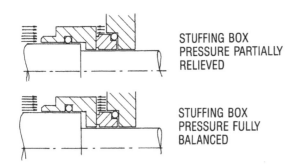

STUFFING BOX PRESSURE PARTIALLY RELIEVED

STUFFING BOX PRESSURE FULLY BALANCED

Fig. 9-23 Principles of hydraulic balance.

A seal incorporating this feature is called a **balanced seal** while one without it is referred to as an **unbalanced seal**. Balancing becomes particulaly important when stuffing box pressures are high and are likely to cause excessive loading of seal faces and hence overheating and wear. A balanced seal requires either a step in the shaft or the inclusion of a shaft sleeve in order to provide the necessary offset.

MATERIALS

A range of materials is used for mechanical seals depending on the conditions under which they operate.

Seal faces are available in various combinations including the following:

    stainless steel/carbon
    lead bronze/carbon
    stellite/carbon
    chrome oxide/carbon
    ceramic/carbon
    tungsten carbide/tungsten carbide
    tungsten carbide/carbon

The carbon used for seal faces is used in composition form usually with either a metal or a resin filler.

Other seal components such as springs, thrust collar, sleees, etc., are commonly made from stainless steel unless special corrosion resistant properties are required, in which case materials such as Hastelloy or even titanium may be used.

Secondary seal materials are selected on the basis of compatibility with the fluid sealed and the temperature conditions. Synthetic rubbers and PTFE are the most common materials used.

TYPES AND ARRANGEMENTS

There is a range of options available in the design of mechanical seals creating a wide variety of possible arrangements. In addition to the choice between a balanced and an unbalanced seal and the range of materials outlined above, the maintenance technician should be aware of the following alternatives available from most manufacturers.

Mechanical loading

There are several methods of providing mechanical loading to the seal faces.

Single coil spring

This arrangement is shown in Fig. 9-20. The advantage of this method is that it eliminates the necessity for drive pins or similar devices to drive the rotary seal ring. By choosing a coil spring of the correct hand, the rotation of the shaft can cause the spring to tighten against the neck of the rotary seal ring at one end, and the thrust collar at the other, as

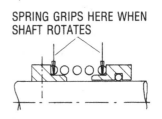

SPRING GRIPS HERE WHEN SHAFT ROTATES

Fig. 9-24 Single coil spring tightens with rotation.

shown in Fig. 9-24, providing a positive drive. This helps to overcome the problem of side thrust and tilting moments associated with drive pins.

Multiple coil spring

A single coil spring can be replaced by a number of smaller springs distributed around the shaft.

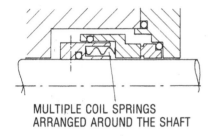

MULTIPLE COIL SPRINGS
ARRANGED AROUND THE SHAFT

Fig. 9-25 Multiple coil spring.

The advantages of such an arrangement are that a more even distribution of loading on the seal faces can be achieved and the length of the seal assembly can be reduced. They are, however, more susceptible to blockage and interference from solids or sludge.

Bellows

The common alternative to the use of springs is to use a set of metal bellows.

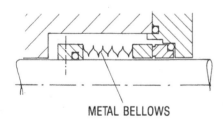

METAL BELLOWS

Fig. 9-26 Metal bellows.

One end of the bellows is welded to the thrust collar which in this case is also machined to accommodate a secondary seal. The other end of the bellows is welded to the rotating seal face which is held clear of the shaft by the rigidity of the bellows and is not required to carry a secondary seal.

The advantage of this type of arrangement is that there is no sliding elastomeric seal under the rotating seal ring, and therefore drag or hang-up on the shaft and also wear of the shaft or sleeve are eliminated.

This feature also enables a bellows seal to operate at higher temperatures than a spring loaded seal.

Bellows made from PTFE can also be used to protect the moving parts of a seal, such as the springs, from contact with the sealed fluid thus preventing corrosion. This arrangement is demonstrated in Fig. 9-27.

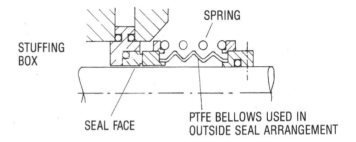

Fig. 9-27 PTFE bellows can protect the moving parts of the seal from contact with the sealed fluid.

Seal location

The most conventional arrangement is to locate the seal inside the stuffing box as shown in Fig. 9-20. It is also possible to locate the seal outside the stuffing box as shown in Fig. 9-28.

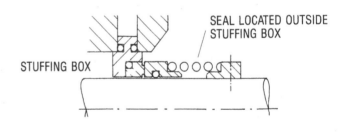

Fig. 9-28 Location of the seal outside the stuffing box.

The advantage of this arrangement is that the sealed fluid is shielded from the moving parts of the seal and corrosion can be eliminated. The major drawback is that the hydraulic pressure now tends to force the seal faces apart. This makes the seal particularly vulnerable to sudden surges of pressure that may open up the seal faces.

Seal combinations

Although the majority of sealing applications can be satisfied by a single seal, in some circumstances a double seal arrangement can be used in order to provide extra protection. This type of arrangement is required when the sealed fluid is not capable of providing the necessary properties to lubricate the seals or when it contains solids or abrasives that could affect the seal faces. The use of a double seal arrangement allows a secondary fluid to be introduced into the space between the two seals. The fluid may be water or a neutral lubricant circulated by an independent system. It is important that the fluid chosen is compatible with the sealed fluid so that contamination can be avoided.

There are a number of ways in which a double seal can be arranged.

Back-to-back

This arrangement consists of two seals mounted in the same stuffing box as shown in Fig. 9-29.

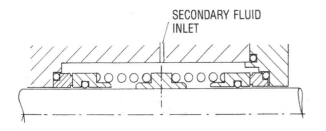

Fig. 9-29 Back-to-back arrangement.

To ensure lubrication of the inboard seal the pressure of the secondary fluid in the stuffing box must be greater than the pressure of the sealed fluid. The example given shows unbalanced seal arrangements but balanced seals can be used if pressure differentials become too great.

Tandem

An alternative to the back-to-back arrangement is the tandem double seal which consists of two seals mounted in the same direction, one inside the stuffing box and the other in a cavity created by a second seal plate as shown in Fig. 9-30.

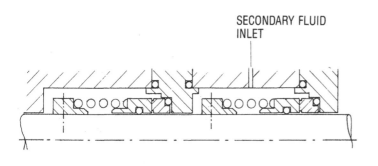

Fig. 9-30 Double seal arranged in tandem.

The inboard seal contains the sealed fluid in the conventional stuffing box but should leakage occur this will be contained by the outer seal which is lubricated by a secondary fluid.

Inside-outside

Where space limitations prevent the use of a tandem seal an arrangement in which a conventional inside seal is backed up by an outside seal can be used. A secondary fluid can be introduced into the cavity between the two sets of seal faces shown in Fig. 9-31.

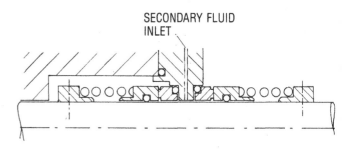

Fig. 9-31 Inside-outside arrangement.

The secondary fluid may be supplied at a higher pressure than the sealed fluid, in which case it lubricates both sets of seal faces. Alternatively, it may be supplied at a lower pressure in which case the outside seal merely acts as a back-up.

Seal and bushing

In situations where the sealed fluid is not hazardous and has adequate lubricating properties but where some control of leakage is required a throttling bush may provide sufficient protection.

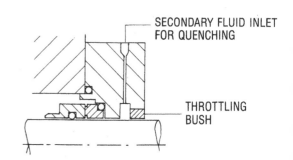

Fig. 9-32 A throttling bush.

This arrangement can also be used to provide quenching to the seal faces by a secondary fluid supplied via the seal plate.

Combination seal arrangements of the types outlined above provide a number of advantages over single seal arrangements:

- They can handle extreme pressure differentials by creating a progressive step-down in pressure across the two seals.
- They provide a buffer zone against the escape of toxic or hazardous fluids.
- They are capable of sealing against a vacuum.
- They are capable of sealing a gas or a liquid which does not possess adequate lubricating properties.
- Fluids containing solids or abrasives can be kept away from the seal faces.
- The hazards associated with the leakage of a high temperature process fluid can be reduced by circulating a lower temperature secondary fluid.

ENVIRONMENTAL CONTROL

An important factor in determining seal performance is the nature of the environmental conditions surrounding the seal face. The factors that affect performance are temperature, corrosion and contamination and these can be controlled by various combinations of flushing and quenching the area around the seal faces.

Flushing

Lubrication and cooling of the seal faces may be accomplished by directing a continuous flow of liquid to the seal interface as shown in Fig. 9-33.

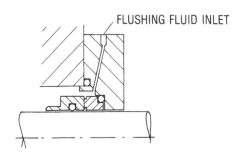

FLUSHING FLUID INLET

Fig. 9-33 A continuous flow of liquid is directed to the seal face.

The liquid used for flushing may be supplied directly from the pump discharge or may be supplied from an external source and is normally directed into the seal area via a tapping in the seal plate or via the lantern ring connection in the stuffing box. A complete flushing system may need to incorporate a cooling device and some means of solids removal such as a cyclone separator.

The advantages of flushing can be summarised as follows:

- Flushing overcomes the tendency of highly volatile fluids to vaporise around the seal interface.
- Fluids that operate close to their melting point can be prevented from crystallising.
- Seal operating temperatures are reduced.
- Flushing prevents solids from accumulating around the seal interface.
- If an external fluid is used for flushing at a pressure above that of the sealed fluid, then the sealed fluid can be eliminated from the stuffing box. This is useful in the handling of corrosive or toxic fluids as long as the flushing liquid is compatible with the sealed fluid.

Quenching

The creation of a buffer zone behind the seal face by the addition of a backing gland or bushing allows a quenching fluid, which is usually supplied from an external source, to be directed into that zone as shown in Fig. 9-32.

The fluid used for quenching should be clean and should not contaminate the interface area.

The advantages of quenching are:

- Leakage of toxic or hazardous fluids can be prevented from reaching the atmosphere.
- Quenching with a fluid above the temperature of the sealed fluid can prevent crystallisation at the seal interface.
- As well as helping to cool the seal interface, quenching minimises the transfer of heat along the shaft to the pump bearings.

MAINTENANCE PRACTICES

General

A number of general considerations should be taken into account when working on mechanical seals.

- Like bearings, mechanical seals are manufactured to fine tolerances and high surface finish and therefore cleanliness is of the utmost importance.
- The key to good seal performance is to ensure that seal faces are square, correctly loaded and properly lubricated.
- A seal should never be run dry, not even for a few seconds.
- Seal rings are lapped together as pairs and a used ring should never be mated with an unused ring.

Assembly

Before assembly is begun the following points should be considered.

- Check the manufacturer's drawings and instructions to ensure that the assembly procedures are correctly understood.
- Make sure that the necessary tools and equipment are available.
- Select a working environment that is clean and dust free and where adequate lighting exists.
- Seal faces should be left in their protective plastic cover until ready for fitting.
- Check that the shaft is clean and free from burrs and other damage and also check that the stuffing box is clean and that flange faces are free from damage.
- To guarantee the optimum operation of the seal, it may be necessary to check shaft run-out, deflection and stuffing box geometry.

The following checks are recommended:

Shaft run-out

Mount dial indicators as shown in Fig. 9-34 to check shaft run-out and whether or not the shaft is bent.

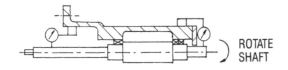

Fig. 9-34 Checking shaft run-out.

Shaft deflection

By lifting the shaft and observing the dial gauge mounted as shown in Fig. 9-35 a check can be made on the amount of shaft deflection that exists due to bearing wear.

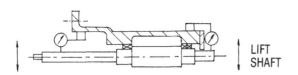

Fig. 9-35 Checking shaft deflection from the dial gauge.

Shaft float

Measure end play by setting up the dial gauge against a shoulder. The measurement recorded should be checked against the manufacturer's recommendations.

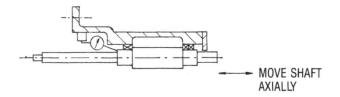

Fig. 9-36 Measuring shaft float.

Sleeve concentricity

If a shaft sleeve is used then this should also be checked for run-out in the same way as the shaft.

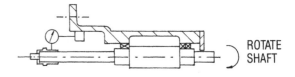

Fig. 9-37 Checking for run-out in the shaft sleeve.

Concentricity of stuffing box bore

A check on the concentricity of the stuffing box bore may be necessary to ensure that the shaft is centralised in the stuffing box. An adapter may need to be made for the indicator and the pump assembled without the seal as shown in Fig. 9-38.

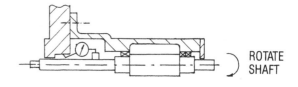

Fig. 9-38 Checking concentricity of the stuffing box bore.

Squareness of stuffing box face

To ensure that seal faces run square a check on the squareness of the stuffing box face can be carried out as shown below. A maximum run-out of 0.075mm (0.003″) is recommended. (Fig. 9-39)

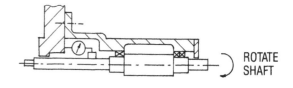

Fig. 9-39 Checking squareness of the stuffing box face.

The assembly procedure itself should be carried out according to the manufacturer's instructions and with consideration given to the following points:

- Shaft shoulders should be chamfered to enable O-rings to be installed without damage.
- If the seal includes a single coil spring make sure that it is the correct hand. The spring should tighten up due to the rotation of the shaft.
- A trace of lubricant may be used to assist the fitting of the rotating seal assembly but no lubricant should be applied to the seal faces.
- To ensure the correct loading of the seal faces the thrust collar must be fastened to the shaft or shaft sleeve in accordance with the dimensions provided by the manufacturer. To provide a reference, a witness mark can be made on the shaft in line with the stuffing box face as shown in Fig. 9-40.

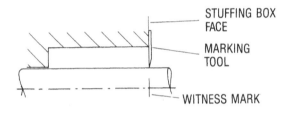

Fig. 9-40 Marking the shaft.

- When fitting the stationary seal ring into the seal plate care should be taken not to twist the ring packing as this may cause damage. A small amount of lubricant on the O-ring will help to push it home square without the need for twisting.
- If flushing connections are provided, and the installation is new or the fluid pumped contains solids, make sure that a strainer is installed in the circulation line to the seal plate.

Start-up

The most important consideration during start-up is that the seal should not start up dry. If the seal is provided with an external flushing connection this must be turned on before the machine is started. Whatever method is used to provide lubrication to the seal faces, steps should be taken to ensure that the stuffing box is flooded before the machine is started. However, only the seal fluid or the

secondary flushing fluid should come into contact with the seal faces. Do no apply any oils or other lubricants to the seal faces.

FAILURE PATTERNS

When a seal begins to leak it is often difficult to tell precisely how it is leaking. There are several possible ways in which this may occur and these are shown in Fig. 9-41.

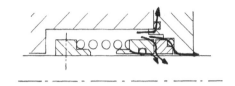

Fig. 9-41 Leakage paths for a mechanical seal.

Leakage of the seal plate gasket will be evident, but it is not so easy to determine whether leakage is occurring across the primary seal faces or the secondary seals.

The majority of seal failures occur due to leakage across the primary faces but careful inspection of all components will often be necessary to make sure that the secondary seals are not to blame.

Seal leakages that occur on start-up are usually due to incorrect assembly such as the fitting of the wrong hand spring, chipped or cracked carbon rings or trapped or damaged O-rings. Inadequate lubrication at start-up may also damage seal faces so that they begin to leak immediately.

There are various symptoms which can be identified by examining the seal components that will assist in determining the cause of failure. The most common causes of seal failure and the effects they produce are as follows.

Vaporisation

When the temperature of the seal interface becomes too great and local boiling occurs, the seal faces pop apart and make a puffing noise.

The stationary carbon face becomes pitted with associated 'comet trailing' and the outside diameter is likely to be chipped. The continual impact of the seal faces as they are blown apart by the vapour causes radial cracks to appear on the rotating face. When the sealed fluid is water, vaporisation may cause the seal faces to blow open and remain open.

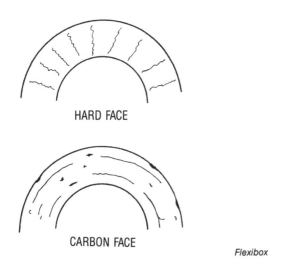

HARD FACE

CARBON FACE

Flexibox

Fig. 9-42 Typical evidence of vaporisation.

Dry running

When the seal faces are inadequately lubricated they are subject to rapid wear in the form of scoring and grooving as shown in Fig. 9-43.

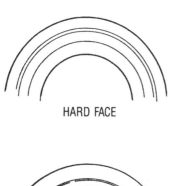

HARD FACE

CARBON FACE

Flexibox

Fig. 9-43 Rapid wear from dry running.

There may also be evidence of overheating in the form of hardening and cracking of the secondary seals if the problem has existed for some time.

Abrasives in sealed fluid

If abrasive matter is present in the sealed fluid this may cause rapid wear of the seal faces and also a solids build-up around the rotating ring secondary seal as shown in Fig. 9-44.

HARD FACE

CARBON FACE

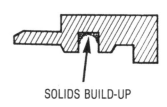

SOLIDS BUILD-UP

Flexibox

Fig. 9-44 Results of abrasive material in sealed fluid.

The ideal solution to this problem is to exclude the solids from the seal faces by flushing with an external source of clean fluid. Alternatively, wear-resistant seal faces may be required (e.g. tungsten carbide).

Sludging and bonding

These two problems produce very much the same kind of seal face damage. Sludging is associated with the sealing of high viscosity liquids and occurs when the shear stresses between the seal faces exceed the rupture strength of the carbon and tear away the surface of the stationary seal ring. This is especially likely to occur when seal temperatures drop and the viscosity of the interface film increases causing problems on start-up. A similar problem occurs when the fluid actually crystallises while the unit is shut down and forms a temporary bond between the faces. When this is pulled apart on start up damage to the carbon face results.

There are a number of possible solutions to these problems depending on the actual circumstances. Assuming that the liquid viscosity lies within the capabilities of the seal then the important considerations are whether there is adequate circulation of fluid around the seal interface area and whether the temperature of the seal can be prevented from falling below the levels at which problems occur. Fluid circulation will depend to a

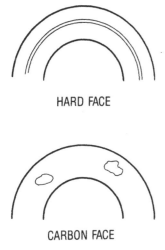

HARD FACE

CARBON FACE

Flexibox

Fig. 9-45 Typical evidence of sludging and bonding.

certain extent on the pressure available in the circulation system and this may need to be boosted if it is not adequate. Some means of preheating the seal before start-up, either by steam tracing, direct injection to the stuffing box or steam circulation through the seal plate may be required to avoid crystallisation.

Distortion

Under some circumstances it is possible for the seal faces to distort. This may be caused by uneven pressure of the drive spring or incorrect assembly. In

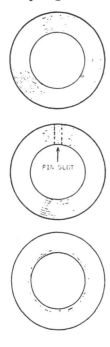

PIN SLOT

Flexibox

Fig. 9-46 Results of distortion.

some cases incorrect storage may also cause the same problem. The results of distortion are uneven running marks on the seal faces as shown in Fig. 9-46.

The problem of distortion may be overcome by re-lapping the seal faces unless the degree of distortion is too great, in which case new seal rings may be required.

Coking

When the sealed fluid is a hydrocarbon at high temperature, any slight, even minute, leakage past the seal tends to carbonise and cause the secondary seal under the rotating ring to jam up. This prevents it from sliding to take up wear at the primary seal face. This is demonstrated in Fig. 9-47.

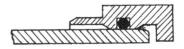

Fig. 9-47 Coking.

The solid build-up of carbonised material on the atmospheric side of the sliding seal eventually jams the rotating seal ring on the shaft or sleeve.

The solution to the coking problem is to provide a steam quench to ensure that the seal temperature stays above that at which coking takes place.

Sleeve damage

It is possible for the sleeve or shaft to become damaged under the rotating ring seal. This, too, prevents the seal ring from sliding along the shaft. The two most likely causes of such damage are vibration and fretting corrosion.

Vibration of the shaft is likely to cause interference between the shaft or sleeve and the underside of the seal ring adjacent to the O-ring groove. This leads to wear and marking of the shaft or sleeve surface as shown in Fig. 9-48.

Flexibox

Fig. 9-48 Wear and marking from vibration of the shaft.

When vibration is very slight fretting may occur under the O-ring itself with consequent damage to the shaft or sleeve as shown in Fig. 9-49.

Flexibox

Fig. 9-49 Fretting damage.

Elimination of the source of vibration is the ideal way to solve such problems but if this cannot be achieved totally then hard facing of the sleeve or shaft will help reduce the effects of the problem.

O-ring damage

Damage to the secondary seals can occur in a number of ways. Apart from errors made during assembly it is usually assocated with the conditions of operation. Overheating of O-rings causes hardening and cracking and may indicate either that the O-ring material is not suitable for the duty or that improved cooling of the seal is required.

Extrusion of the O-ring, shown in Fig. 9-50, may be caused by excessive pressure or incorrect clearances.

If the O-ring material is incompatible with the seal fluid then it may appear to be eaten away as shown in Fig. 9-51. The seal manufacturer should be consulted to consider the selection of the seal materials.

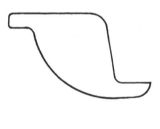

Fig. 9-50 Extrusion of the O-ring.

Flexibox

Fig. 9-51 Incompatibility of O-ring material and seal fluid.

Carbon ring erosion

Erosion damage to the carbon ring, shown in Fig. 9.52, may occur due to either the circulation pressure being too high or the presence of abrasive particles in the circulation flow.

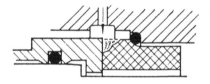

Flexibox

Fig. 9-52 Erosion damage to the carbon ring.

9.5 PACKINGS

The use of packings is one of the oldest approaches to solving dynamic sealing problems. Rings of soft packing material are usually fitted into a stuffing box around the shaft and compressed by a gland ring which forces the packing into contact with the shaft and the stuffing box bore. This method of sealing is also used as a static seal and for reciprocating mechanisms.

SEALING PRINCIPLES

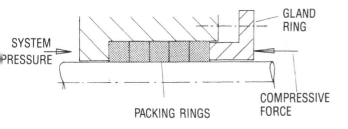

Fig. 9-53 The operating principles of a packed gland.

There are various forms that the packing may take, the most usual being rings of soft woven or bonded material that are cut to length and assembled around the shaft inside the stuffing box. A gland ring or follower is pulled up against the stuffing box casing to compress the rings and cause them to expand radially. This radial expansion brings the packing into loaded contact with both the shaft and the stuffing box bore to create the seal. One of the factors affecting the efficiency of the packing is the number of rings. It has been demonstrated by experience that the optimum number of rings is five, due to the manner in which the pressure varies over the packing area. As can be seen from Fig. 9-54, if there are too many rings, the innermost do very little work.

The materials used for packing possess self-lubricating properties which protect the shaft from wear, at least on start-up. However, this lubricant would soon leach out and the packing overheat if it were not supplemented in some way.

In cases where the pump suction pressure is above atmospheric pressure the sealed fluid can be used directly to lubricate the packing by allowing a small amount of controlled leakage through the gland. If the pump suction pressure is below atmospheric pressure, liquid from the pump discharge may be directed into the gland area via a lantern ring which is installed between the rows of packing. If the sealed

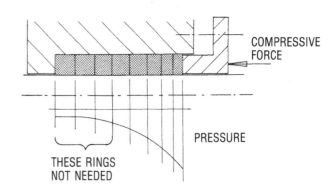

Fig. 9-54 If there are too many rings, the innermost do little work.

fluid contains abrasives or is unsuitable for lubrication for some reason a secondary fluid may be directed to the gland area via the lantern ring. These three alternatives are shown in Fig. 9-55.

Whatever arrangement is used it is imperative that the packing is not allowed to run dry. A properly adjusted packed gland should exhibit a slow but steady leakage of liquid from the gland ring.

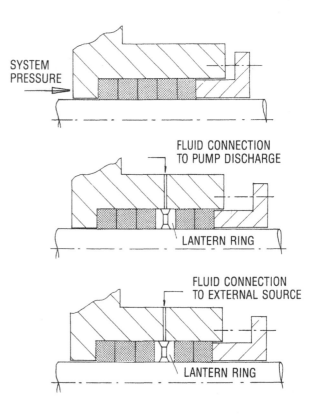

Fig. 9-55 Alternative methods of lubricating a packed gland.

MATERIALS

A wide variety of packing materials is available, the traditional type based on a lubricated fibre yarn remaining the most common. The yarns used may be twisted, plaited or braided into square or round sections using a variety of patterns.

Traditional yarn materials such as hemp, flax, cotton and asbestos (asbestos is no longer used as widely because of the associated health hazards) are still in common use and to these have been added a range of synthetic materials including nylon, rayon and, more recently, PTFE and aramid fibres. Dry lubricants, such as graphite and molybdenum disulphide, are generally more popular than oils and greases while the low friction properties of PTFE have led to its increasing use as a lubricant also. The combination of aramid fibres, which have high tensile strength and heat resistance, and PTFE lubricant has produced a new generation of packing materials which give long life and high chemical resistance.

The other significant area of development in packing materials has been in the use of graphite. Graphite filament yarns have been developed that can operate at extremes of temperature, pressure and speed in the most aggressive of chemical environments. The use of exfoliated graphite foil represents a valuable innovation which, although expensive compared with traditional packings, has proved highly successful in many applications.

Metallic foils are sometimes used for extremely high temperature applications with aluminium, copper and lead lubricated with graphite being the most common.

MAINTENANCE PRACTICES

The performance of compression packings is very much dependent on the procedures that are followed during installation and start-up. The following steps are recommended.

1 Make sure that the stuffing box is clean and free from old packing material.
2 Check the shaft for run-out and deflection as recommended for mechanical seals in Section 9.4.
3 Check the shaft for wear and scoring in the gland area. Replace or build up and remachine if necessary.
4 Make sure that the correct size and type of packing is available. If necessary, check the shaft o.d. and stuffing box i.d.
5 If conventional packing supplied in a continuous length is to be used, then cut it into separate

rings. Never wind a coil into the stuffing box. Cut the rings on a mandrel or exposed section of shaft as shown in Fig. 9-56.

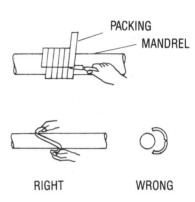

Fig. 9-56 Cut packing into rings.

Cut each ring square as shown and twist the rings to remove them from the mandrel to avoid damaging the inside diamater.
6 Install the rings one at a time and use a tamping tool to make sure that each is properly seated. A light coating of clear oil on the shaft and stuffing box bore may assist installation. Stagger the joints by at least 90° for each successive ring.
7 Install the lantern ring (if included) so that it is in line with the flushing fluid inlet. Remember that the lantern ring will move down the stuffing box when the follower is tightened and so many need to start off slightly behind the fluid inlet.
8 Install the gland ring or follower and pull up finger tight.
9 If a secondary flush is to be used, open the inlet to the gland and ensure that fluid is present.
10 Start up the machine and allow the gland to leak freely for a minute or two to ensure that the packing is well lubricated. Take up the gland ring bolts until leakage is reduced to a tolerable level.
11 After about one hour take up the gland until the leakage rate is around 1-2 drops per second for a ∅25mm (1″) shaft, more for larger sizes.
12 Check the gland periodically to adjust the leakage rate and check the stuffing box temperature. If the gland runs hot increase the leakage rate. Never, for any reason, allow the gland to run without leakage as this will quickly cause over-heating and destroy the packing.

Exfoliated graphite

Unlike most packing materials exfoliated graphite is supplied as ribbon or tape which can be used to form packing rings *in situ* to suit any sized gland. The principle on which these are prepared is shown in Fig. 9-57.

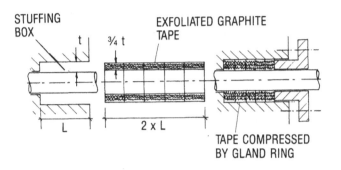

Fig. 9-57 Forming packing rings from exfoliated graphite

The tape or ribbon is wound on to the shaft to form a ring of approximately three-quarters of the thickness of the gland space as shown. As the compression required of the material is normally around 50%, a sufficient number of rings to form a length approximately twice that of the gland will be required. Hence the tape width should be around two-fifths of the final packing depth if five rings are to be used. When all five rings have been built up and slid into the stuffing box the gland ring can be installed and tightened to give the rings compression. As with conventional packings, a slight leakage should be maintained to ensure that the package is lubricated and gives maximum life.

9.6 LIP SEALS

Radial lip seals are used primarily to retain oil and other lubricants in equipment operating with rotating shafts. They are also used to exclude dirt and other contaminants. They are generally only suitable for sealing against low pressures of the order of 35 kPa (5 psi).

SEALING PRINCIPLES

A radial lip seal consists of an elastomeric sealing ring contained in a metal case as shown in Fig. 9-58.

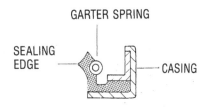

Fig. 9-58 A radial lip seal.

The sealing ring is formed with a knife edge at the contact point which engages with the rotating shaft. A garter spring may be fitted inside the ring to help load the seal. The flexible element may be bonded to the case or the parts may be made separately and then rolled or crimped together.

Sealing occurs by establishing a load at the point of contact between the sealing ring and the shaft. The preload created by the interference fit of the elastomeric ring is supplemented where necessary by the mechanical pressure of the garter spring. A hydrodynamic oil film is formed at the point of contact and should be around 0.025mm (0.001″) thick to prevent friction and wear. If the film is any thicker then leakage tends to occur.

The action of the seal relies on the continuous presence of a clean oil film to protect the sealing edge of the elastomeric ring. The shaft surface should be machined to a high order and yet remain rough enough to promote oil retention. A surface finish of the order of 0.025−0.050mm (0.001″−0.002″) is recommended.

TYPES AND MATERIALS

A radial lip seal may be of bonded construction in which the flexible sealing element is permanently bonded to the casing or it may be assembled from separate components which are crimped together as shown in Fig. 9-58. The principles of bonded construction are shown below in Fig. 9-59.

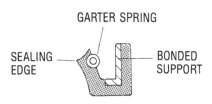

Fig. 9-59 The principles of bonded construction.

The seal may have a single or double lip arrangement with either or both lips being spring loaded. Typical alternatives are shown in Fig. 9-60.

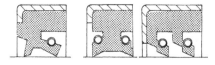

Fig. 9-60 Typical lip seal arrangement.

The second lip of a double lip seal performs the function of a dirt seal or, in a back-to-back arrangement, provides sealing in both axial directions.

In order to increase the sealing efficiency the atmospheric side of the seal lip may have ribs or grooves moulded into it. These create a hydrodynamic pumping action during operation so that any fluid that leaks past the lip is pumped back into the contact area. This feature allows the seal to

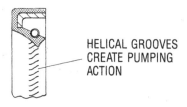

Fig. 9-61 Moulded grooves in the atmospheric side of the seal lip help to prevent leakage.

operate at lower lip pressure and hence reduces friction and wear. An example is shown in Fig. 9-61.

Different materials are used for the flexible element of a lip seal depending on the operating conditions. These include nitrile, polyacrylate, fluorocarbon, polyeurethane, PTFE, silicone and leather. The most common of these is nitrile which is compatible with most lubricants and suitable for temperatures up to 130°C (270°F).

MAINTENANCE PRACTICES

One of the most common causes of lip seal failure is damage that occurs during the handling or fitting of the seal. Hence great care should be taken during assembly and the following procedures and precautions are recommended.

1 Before attempting to install a lip seal, make the following checks:
   Check the seal dimensions against the shaft and housing and make sure they match specifications.
   Check the seal lip for damage.
   Check the shaft surface finish and make sure it is free of nicks and burrs.
   Make sure the seal is clean and that the garter spring is properly located.
2 Make sure that the seal is orientated with the lip facing towards the lubricant.
3 Apply a coating of lubricant to the sealing edge and to the shaft to aid installation.
4 If the seal is metal-cased then a coating of gasket cement around the outside may be required to prevent the seal from leakage around the housing.
5 Use a properly designed tool to push the seal home into the housing.

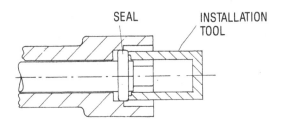

Fig. 9-62 Use the correct tool to push the seal home into the housing.

If the correct tool is not available use a suitable ring that contacts the seal casing around the outside diameter. Never press against the sealing lip.

6 If the seal has to be fitted over a keyway, splines or other sharp corners use a protective cone or sleeve as shown in Fig. 9-63.

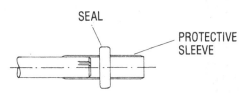

Fig. 9-63 Using a protective sleeve.

7 Make sure that the seal is not cocked in the housing and has been pushed home firmly all round.
8 Turn the shaft by hand to make sure it runs freely.
9 The seal may leak slightly on start up until the lip wears in and the hydrodynamic film is established. Once the seal is seated it should run without any measurable leakage.

FAILURE PATTERNS

The most common reason for failure of a radial lip seal is mechanical damage to the sealing edge. The dangers of causing damage during assembly have already been pointed out. They should be avoided if correct procedures are followed. The condition of the shaft is critical and if a seal does not give satisfactory performance then the shaft surface should be carefully inspected for imperfections.

Running conditions can also affect seal performance. The operating speed and temperature must be kept within the design limits of the seal or early failure due to excessive wear can be expected. Inadequate lubrication will also lead to rapid wear and early failure and it should be recognised that operating speeds have an effect on the characteristics of the lubricant. As the speed increases the operating temperature rises and causes the viscosity of the lubricant to decrease and the lubricant film to thin out. This causes increased friction and a further rise in temperature and this whole cycle greatly reduces the life of the seal.

In addition to the above considerations, care should be taken to ensure that the seal is installed the right way round. If hydrodynamic ribs are provided the direction of rotation may be critical and should be checked.

9.7 CIRCUMFERENTIAL SPLIT RING SEALS

Circumferential split ring seals, sometimes referred to as carbon ring seals, are a modified version of a piston ring seal and are designed to seal gases and vapours at high temperatures. They are often found in steam turbines.

SEALING PRINCIPLES

In their simplest form, split ring seals consist of three or more carbon ring segments cut radially to form a set as shown in Fig. 9-64.

CARBON RING SEGMENTS

Fig. 9-64 Simplest form of split ring seal.

The rings are held together by a garter spring which fits into a machined groove around the outside of the ring segments as shown in Fig. 9-65. A stop or pin prevents the rings from rotating inside a slot in the machine housing.

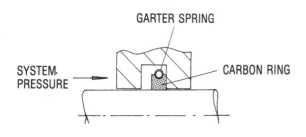

GARTER SPRING

SYSTEM PRESSURE

CARBON RING

Fig. 9-65 Typical split ring seal arrangement.

The clearance between the ring and the shaft is sufficiently small to prevent leakage and the internal pressure forces the ring segments against the side of the slot to create an axial seal in a similar way to a mechanical end face seal. Diametral clearances for carbon ring seals usually run around 0.0125mm −0.0625mm (0.0005″−0.0025″).

The seal is free to move radially in the housing and can therefore compensate for small irregularities in shaft movement without allowing excessive leakage.

Axial movement of the shaft can also be tolerated by this type of seal.

TYPES AND ARRANGEMENTS

Carbon ring seals are usually installed in sets which are spread axially along the shaft as shown in Fig. 9-66.

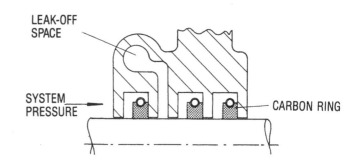

LEAK-OFF SPACE

SYSTEM PRESSURE

CARBON RING

Fig. 9-66 Carbon ring seal set.

A leak-off space is often provided so that any fluid that leaks past the inner ring is allowed to blow off freely and not pass across the outer rings which usually protect a bearing.

More complicated split ring seal arrangements employ up to three sealing ring elements with each consisting of several segments. Some of the rings have joints cut tangentially and are self-compensating for wear. Two primary rings with staggered joints provide the seal against the housing and are backed up by a secondary ring which covers the radial joints in the primary rings. Both sets are loaded by garter springs as before and a typical arrangement is shown below in Fig. 9-67.

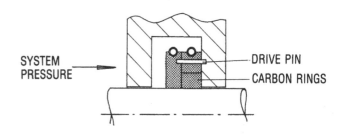

SYSTEM PRESSURE

DRIVE PIN

CARBON RINGS

Fig. 9-67 A more complicated split ring seal arrangement.

MAINTENANCE PRACTICES

Carbon ring seals generally require little maintenance if the shaft is running true and the housing is clean and free from damage to the sealing surfaces.

The following points should be considered.

- There is no particular tolerance on carbon ring clearances. Seals should be replaced or repaired when leakage becomes excessive.
- When a machine is shut down, the carbon ring seals should be inspected for wear and sticking. Debris may build up in the seal slot and prevent the sealing ring from moving freely.
- Spring tension should be checked and springs should be replaced if necessary.
- Rings should be dismantled carefully, cleaned, and inspected for damage.

- Rings should always be marked so that they can be reassembled in the same positions.
- Worn rings can be refitted by scraping if necessary. The inner surface should be scraped first to fit the shaft and then the face. The ends of the segments should be scraped last until the complete ring fits the shaft but does not grip it when the garter spring is installed.
- Broken or damaged rings should always be discarded and replaced.
- Before reassembling the seal, the shaft and seal slot face should be carefully examined for damage and build-up of debris.
- Always handle carbon rings with great care because of their fragility.
- When refitting rings make sure that no dust or dirt is trapped against the sealing faces or between the ends of the segments.

9.8 FELT SEALS

Felt seals are primarily used as oil and dust seals and provide a cheaper alternative to radial lip seals under operating conditions that are not too severe.

SEALING PRINCIPLES

Felt is a fabric composed of interlocking wool, or other animal, vegetable or synthetic fibres. When used for sealing purposes it is normally presaturated with a lubricant of slightly higher viscosity than that being sealed. The seal is usually cut from felt sheet and is held in a casing of which there are various designs.

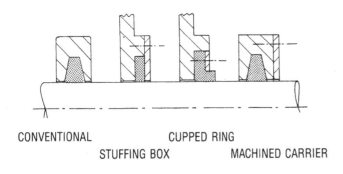

CONVENTIONAL   CUPPED RING

STUFFING BOX   MACHINED CARRIER

Fig. 9-68 Examples of felt seal designs.

The configuration of the felt ring should be such that the difference between the inner and outer diameter is larger than the thickness and it is usual for the thickness to be greater at the i.d. than the o.d. The seal is normally supplied in the form of a cartridge which is designed to fit a machined housing in the bearing assembly.

The interference between the seal and the shaft should be only slight and contact pressure should be low.

Felt seals are not suitable for operation with low viscosity lubricants or for sealing against pressure.

MAINTENANCE PRACTICES

There are a number of considerations that should be kept in mind when handling felt seals.

- Felt is are relatively delicate material and can be easily damaged if it is unduly compressed or stretched.
- Like other types of seals the condition of the shaft at the seal interface will play an important part in the effectiveness of the seal.
- It is normal practice to replace a felt seal, regardless of its condition, whenever a machine is overhauled.

9.9 CLEARANCE SEALS

All the seals discussed so far have been contact seals which rely on direct lubricated contact between a sealing element and the rotating shaft. Clearance seals operate without contact and rely on a small clearance gap to create a pressure drop between the system pressure and the atmosphere.

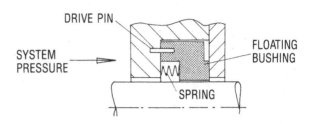

Fig. 9-70 A floating bushing may be spring loaded.

BUSHINGS

The simplest type of clearance seal is a bushing which operates because of the throttling action provided by a small clearance gap over the length of the bushing.

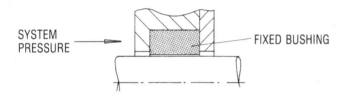

Fig. 9-69 A typical example of a bushing.

The long but relatively small clearance space between the shaft and the bushing creates a restriction to flow which will vary depending on the viscosity of the fluid. Hence bushing seals are not as effective with low viscosity liquids and are no use for sealing gases and vapours.

Types and materials

The simplest type is a fixed bushing of the type shown in Fig. 9-69. The disadvantage of a fixed bushing is that it cannot tolerate any degree of misalignment or eccentricity of the shaft. The only way a fixed bushing can cope with any form of shaft run-out is by increasing the clearance and accepting a higher leakage rate.

To overcome the problem of shaft run-out, a floating bushing may be used. In this arrangement the bushing is allowed to float in the housing and one end is lapped and forms an axial seal against the end of the housing. The bushing is held by a dowel pin so that it does not rotate and may be spring loaded as shown in Fig. 9-70.

For sealing against high pressures, which tend to cause the bushing to stick to the housing, the effect of pressure can be relieved by providing hydraulic

balance in the same way that axial end-face seals are balanced. The end of the bushing is stepped to provide back pressure as shown in Fig. 9-71.

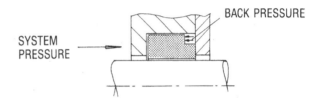

Fig. 9-71 A pressure balanced bushing.

An alternative way to handle high pressure is to replace a single bushing by a series of seal rings, each of which carries only a portion of the total pressure drop. Each ring is free to float in the housing as shown in Fig. 9-72.

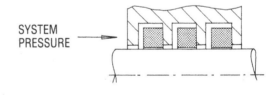

Fig. 9-72 A single bushing can be replaced by a series of seal rings.

Because contact between the shaft and bushing cannot be prevented with certainty it is usual to use a low-friction material for the bushing. Carbon, molybdenum disulphide and PTFE are commonly used, either independently or in composite form, and whitemetals can be used for low temperature service. Higher temperature applications may require the use of aluminium alloys or bronzes.

Maintenance practices

For effective operation, bushings must present the correct clearance gap. Shaft o.d. and bushing i.d. are critical dimensions that should be checked to ensure that the bushing operates effectively. Shaft surfaces should also be inspected for nicks and burrs and excessive machining marks.

Housings should be kept clean and free of debris and bushings should be free to move radially without interference. The ends of floating bushings should be lapped flat and should be free of chips or cracks so that a good axial seal is made with the housing.

Despite the flexibility of a floating bushing, excessive run-out or shaft deflection cannot be accommodated by a bushing seal and should be eliminated.

LABYRINTH SEALS

A labyrinth seal presents a tortuous leakage path to a sealed fluid by providing a series of barriers which dissipate the pressure energy of the fluid in steps. They are used for sealing compressible fluids such as gases and vapours. They can operate at high speeds and temperatures and are often used in high speed machinery such as centrifugal compressors and steam turbines.

Sealing principles

Labyrinth seals can be made in many different forms but the simplest arrangement is the type shown in Fig. 9-73.

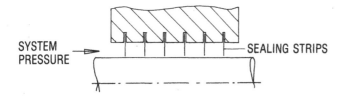

Fig. 9-73 The simplest arrangement of a labyrinth seal.

As the sealed fluid passes through each clearance gap it first accelerates, then decelerates and expands into the following chamber producing turbulence and friction. This results in a loss of pressure energy at each stage with the total pressure differential being spread across the full length of the seal.

The barriers to flow are formed by thin parallel metal strips that are held perpendicular to the shaft and maintain close clearance with the shaft surface. Each chamber has to be wide enough to prevent the

fluid from passing directly through from one clearance gap to the next. With a straight through design of the type shown in Fig. 9-73 some carry-over of kinetic energy from one chamber to another is inevitable but this can be eliminated by modifying the seal geometry and introducing a staggered or stepped arrangement as shown in Fig. 9-74.

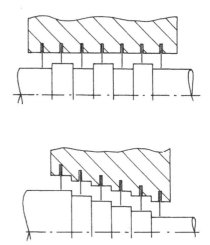

Fig. 9-74 Staggered and stepped arrangement.

Types and materials

Labyrinth seals are usually custom designed to suit each individual machine and materials are selected according to the conditions of operation.

An additional feature that is sometimes desirable involves the provision of a neutral gas purge. The gas is injected into the seal near the high pressure end and then vented to atmosphere either directly or via an eductor system as shown in Fig. 9-75.

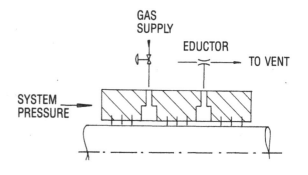

Fig. 9-75 A neutral gas purge.

Maintenance practices

A labyrinth seal should require very little maintenance as long as other machine elements are functioning properly. The clearance gap is vital to the operation of the seal and will depend on the accuracy of machining, concentricity of the housing and the true running of the shaft.

Excessive bearing clearances or any other source of run-out in the shaft will affect seal clearances and may ultimately lead to damage and leakage. Thermal expansion must be allowed for in the design of the seal and changes in the operating temperatures of the machine may affect seal operation. Rigidity of the housing or machine casing is also an important factor in maintaining the correct seal clearances.

When a machine is opened for inspection the sealing tips should be checked for damage and the shaft surface should be checked for interference. If the shaft runs true in the housing and temperature effects are properly allowed for, a labyrinth seal should give unlimited life.

CONDITION MONITORING

The advancement of high technology, especially in the aero-space and nuclear power industries, where safety and reliability are of prime importance, has led to the development of a range of techniques designed to monitor machine operation and generate information that can be used to anticipate breakdown. Monitoring may be carried out on a continuous or an intermittent basis and in some cases can be used to activate a shutdown mechanism.

All condition monitoring techniques rely on monitoring some physical characteristic that reflects the condition of the machine. A normal running level for that characteristic is established when the machine is in good condition and then any significant deviation from that level gives warning that a fault may be developing. This enables a potential fault to be detected before it becomes serious enough to substantially affect machine performance and in time for corrective action to be taken.

These methods are often referred to under the heading of 'Predictive Maintenance' and have been widely adopted in most industries over the last ten years.

Advantages of condition monitoring

Although some of the techniques described in this chapter can be expensive to install, for plant and equipment where there is a high safety and reliability requirement there are distinct benefits to be gained.

- Safety is improved by avoiding the development of dangerous situations that may be hazardous to personnel and other plant and equipment.
- Disruption of production schedules can be reduced by preventing unexpected breakdowns.
- By detecting faults before serious failure occurs, damage to plant and equipment is reduced.
- Condition monitoring eliminates the need to strip down equipment during annual overhaul. As well

as indicating when a machine is becoming unserviceable, monitoring also confirms that a machine is operating satisfactorily.
- The equipment can also be used to assist in troubleshooting and to help identify failed components when breakdown does occur.

METHODS

All the methods described below are designed to monitor the condition of on-line machinery and all can be considered as improved, mechanised versions of inspection by personnel. Before the development of these techniques, condition monitoring was carried out by the operator of a machine who observed the running condition using the senses of sight, hearing, touch and smell. This meant that the operator would **look** for obvious signs of defects, **listen** for unusual sounds, **touch** to check overheating and **smell** to detect burning or overheating. This method of inspection can work quite effectively when the operator is experienced and in constant attendance on the machine. In modern industry however, the majority of machines run unattended and although inspection by personnel is still an important element in the monitoring process, mechanised techniques are essential if constant surveillance is to be achieved.

Temperature monitoring

This is one of the simplest methods available and involves the use of thermocouples or resistance thermometers to measure bearing temperatures. The sensing element should be located within 1.25 mm (0.05″) of the bearing surface and good thermal contact is vital.

Bearing temperatures tend to fluctuate with load so any preset warning levels must take account of normal maximum temperature. Warning levels should be based on rise above normal rather than an arbitrary maximum.

This method can give a warning of several hours before breakdown occurs and can be used to raise an alarm and shut down the machine.

Spectrographic oil analysis

Samples of lubricating oil can be analysed using a spectrometer and the proportions of metal elements present can be determined. The oil samples used must be representative of the total contents of the system and should be taken under normal operating conditions. The major source of metals found in lubricating oil is wear debris, and a knowledge of the component materials can indicate the origin of the metals detected.

This method requires regular sampling so that trends can be established. Normal wear will produce a slow but steady increase in metal levels and it is when a sudden increase is recorded that some malfunction is indicated. It should be remembered when interpreting results that metal levels will be high during a running-in period and will also be affected when oil levels are topped up.

Slow deterioration in operating condition can be detected by carefully monitoring trends but failures which occur rapidly may not be detected by this technique because of the time delay between sampling and obtaining the result of the analysis.

This method, because it is based on an intermittent sampling process, is not suitable for automatic alarm and shutdown.

Particle retrieval

As with the last method, this technique analyses the wear debris present in the lubricating oil but instead of measuring metal levels it is more concerned with the physical size of the particles. Wear debris can be collected by installing magnetic plugs in the system downstream of bearings to catch ferrous particles and by backwashing the oil filter element to collect non-ferrous particles. Debris is inspected and classified with the aid of a microscope.

As with the previous technique, this also relies on observing deviations from a trend established by regular sampling and analysis. During run-in large quantities of very small particles can be expected while normal operation would be expected to yield a smaller quantity of wear particles generally below 25×10^{-6} mm ($1 \times 10^{-6}''$) in size. A steady build up of particles larger than 0.25mm (0.01″) over three or more samples indicates that wear is increasing at a rapid rate and failure may be imminent.

Again, because this method is based on

intermittent sampling, it is not suitable for automatic alarm and shutdown.

Noise monitoring

The noise produced by a machine has traditionally been used as one of the key sources of information about its operating condition. It is still common practice for a technician to hold a screwdriver to a bearing housing in order to detect a malfunction. This simple process can now be systematised using a microphone, amplifier, meter, and continuous recording equipment.

This method is very similar to vibration monitoring which is described below but is not as useful because of the greater difficulty in interpreting results. Hence it is not widely used.

Relative displacement measurement

This method involves the use of devices called **proximity probes** which measure the distance between the probe and a surface such as a rotating shaft. When two probes are set up to view a shaft at right angles, as shown in Fig. 10-1, the outputs can be fed into the x and y channels of an oscilloscope and a display of the shaft locus (the path of movement of the shaft axis) can be seen.

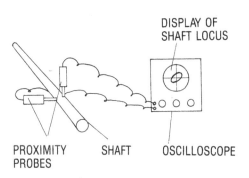

Fig. 10-1 Setting up proximity probes.

Excessive shaft movement can be detected by this method and particular problems, such as shaft whirl and oil whirl, can be analysed using the oscilloscope.

Monitoring is continuous and can be used for automatic alarm and shutdown.

Vibration monitoring

The most widely adopted form of condition monitoring over the last ten to fifteen years involves the recording and analysis of machine vibration patterns. Excessive vibration is a reliable indicator of malfunction and by analysing vibration levels at different frequencies accurate diagnosis of the source of vibration is possible.

Vibration is usually measured by a transducer probe which is attached to a suitable point on the machine and is linked by cable to a remote monitoring unit. The probe, which contains a piezo electric accelerometer, may be permanently installed for continuous monitoring or may be hand held and attached to a portable unit for intermittent checking.

The monitoring equipment may be tuned to record vibration levels at all frequencies or, by the use of filters, may be tuned to a selected frequency band. It is usually especially important to monitor vibration levels at motor drive speed and multiples of drive speed.

This method is ideal for continuous monitoring and vibration sensors can be used to initiate automatic alarm and shutdown devices.

As with all other monitoring techniques this method relies on measuring a change from the pattern associated with normal operation. Vibration monitoring allows a signature pattern to be recorded that represents the vibration levels at all frequencies. An initial trace, made with the machine in good operating condition, will represent the base pattern from which deviations can be observed.

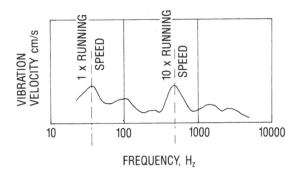

Fig. 10-2 A typical example of a base signature pattern.

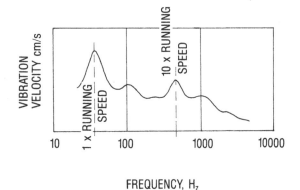

FREQUENCY, H$_z$

Fig. 10-3 A later trace taken for the machine in Fig. 10-2.

The trace shown represents a pump with ten vanes to the impeller and it should be noted that high vibration levels are indicated at running speed and ten times running speed.

A later trace taken for the machine is shown in Fig. 10-3. It can be clearly seen that there has been a marked increase in the vibration level at running speed which might be caused by a loss of balance.

Different mechanical problems give rise to high vibration levels at different frequencies and hence it is possible to use the signature analysis to predict the potential causes of the problem. Some examples are given below.

Source of vibration	Dominant Frequency
Oil whirl	½ x shaft speed
Gear whine	Tooth contact frequency
Bearing defects	Ball or roller speed
Misalignment	2 x shaft speed
Out of balance	Shaft speed
Fan whine	Blade passage frequency

Experience plays a major part in determining the acceptable levels of vibration for particular machines and these will be influenced by the size and type of machine. As a guide to acceptable levels the following chart gives recommended limits for different categories of machine.

Vibration velocity upper limit mm/s rms	Machine class					
	1	2	3	4	5	6
0.28						
0.45			GOOD			
0.71						
1.12						
1.8						
2.8		SATISFACTORY				
4.5						
7.1		IMPROVEMENT				
11.2			DESIRABLE			
18.0						
28.0						
45.0		UNACCEPTABLE				
71.0						

1 Small machines such as electric motors up to 15 kW (20 hp).
2 Medium sized machines up to 75 kW (100 hp) without special foundations or rigidly mounted machines up to 300 kW (400 hp) on special foundations.
3 Large machines mounted on heavy foundations that are rigid in the direction of vibration measurement.
4 Large machines on foundations that are soft in the direct of vibration measurement.
5 Machines with unbalanceable parts (e.g. reciprocating mechanisms) on foundations that are rigid in the direction of vibration measurement.
6 Same as 5, but with soft foundations.

TROUBLESHOOTING

Troubleshooting skills are vital to a maintenance technician. Before effective repairs can be carried out the precise cause of breakdown must be determined and although in some cases this may be clearly apparent, in others it will not and a process of investigation will be required in order for the cause to be found. Although every problem is different in some way, the process of troubleshooting, or problem solving, is essentially the same and can be examined in terms of critical elements that apply in all instances. An understanding of these elements will improve the technician's ability to troubleshoot.

Troubleshooting can be described as a logical system of investigation designed to yield the correct cause of breakdown in the shortest possible time and with the least likelihood of error. The term, **breakdown** is used to indicate any machine condition that is considered to be less than satisfactory according to these factors:

    performance
    downtime
    service life
    efficiency
    safety
    environmental impact
    cost

which have been described in Chapter 1. Hence breakdown refers not just to dramatic failures that render a machine totally inoperable, but also to failures that lead to an unacceptable reduction in performance. This point is emphasised in order to demonstrate that troubleshooting is not some sort of secret weapon brought out of mothballs on special occasions and used only by experts. Whenever a machine fails to meet the criteria of satisfactory operation then the process of troubleshooting must be employed to determine why. It is therefore just as important for maintenance technicians to understand how the process can be most effectively carried out as for other members of the organisation.

APPROACH TO TROUBLESHOOTING

Troubleshooting, like most skills, can be learned, and it is not necessary to assume that it is something for which only certain talented people are qualified.

One of the biggest obstacles to problem solving for many people is developing the confidence to make judgments and then to back them up. Confidence, however, does not just appear out of thin air but is dependent on other factors. It is by understanding and responding to those factors that confidence can be raised and troubleshooting ability improved.

The ability to make successful judgments and become confident in solving problems is based on the following factors.

Knowledge

In order to be able to assess the condition of rotating machinery and to successfully diagnose the cause of breakdown, it is absolutely vital that technicians have a thorough knowledge and understanding of the physical characteristics of the machine and its construction, the principles upon which the machine operates and the function that it performs. If it is not clear to someone how a machine works then it is unlikely that they will have the ability to diagnose the cause of malfunction when it occurs. The greater the knowledge and understanding of the principles of operation, characteristics and function of an item of rotating machinery, then the greater will be the confidence to make judgments about its condition and the greater the likelihood of those judgments being correct.

This does not mean that it is necessary to possess an encyclopaedic amount of knowledge to be successful at troubleshooting. A certain amount of background knowledge is of course expected, and the purpose of this book is to set out some of that information with respect to key machine elements, but particular problems often require specific research to provide the detailed information necessary to solve the problem. When the occasion arises, the technician should know where to find the relevant information, and how to apply it to the problem. For example, it is always wise, before starting any maintenance task, to consult the description of operation and maintenance instructions supplied by the manufacturer. In most cases this will be sufficient to provide the detailed knowledge required although particularly difficult problems may require deeper research.

Logic

A systematic approach is essential in troubleshooting if blind alleys and false conclusions are to be avoided and causes are to be established in the most efficient manner. Troubleshooting can be thought of as a process of elimination. There may be many possible causes of a particular breakdown and troubleshooting is the process by which all but the actual cause are eliminated. Finding the shortest route from a situation of countless possibilities to the real cause of the problem requires a logical, step-by-step approach.

There is no need to be frightened by the word, logic. It is merely the process of making reasonable deductions from existing information, e.g., hammer-blow + thumb-in-line = pain. The ability to apply logic in this way is really the ability to use **common sense** based on knowledge and experience.

Another way to consider the application of logic is to think of it as a question of **organisation**. When planning any activity there is a sensible (i.e. logical) sequence of events that will give the best result. When wallpapering a room, for example, the room is measured before the wallpaper is bought, not after. Logical troubleshooting is merely an extension of this simple principle. All the steps in the troubleshooting process should bring a solution closer by adding to the understanding of the problem and eliminating irrelevant possibilities.

In most cases the correct solution can usually be selected from a range of obvious alternatives by using informally, the logical principles of an organised approach and common sense. Where obvious solutions fail, however, a more formal and systematic approach is called for. The most formal way in which logic can be applied to a problem is through a systematic diagnostic approach of the kind explained on page 128. This more rigorous type of approach is not usually used unless the problem proves to be especially difficult to solve. This approach requires care and a considerable amount of discipline as cutting corners can easily destroy the benefits of such an approach.

Whether logic is applied informally in the form of common sense or formally through a systematised method it can be thought of as the thread that draws the pieces of the puzzle together to produce the right solution. As with a jigsaw, there is only one way the pieces will fit together to make the picture and similarly with a maintenance problem the correct solution can only be found if the pieces of information and evidence are put together in the right way. Logic is the key to achieving this in the most efficient manner.

Experience

Although much can be accomplished with sound technical knowledge and a rational approach, there is little doubt that experience can be invaluable when it comes to troubleshooting. Unfortunately machinery does not always react according to prediction and although a logical explanation is usually possible once the cause of a problem has been found, logical deduction of the cause from the evidence available may not be so simple. Someone with experience should find it possible to make educated guesses and connect cause and effect, whereas this may be difficult for an inexperienced person.

This does not necessarily mean that only the most experienced should turn their hand to troubleshooting although clearly, all other things being equal, an experienced person may have something extra to offer. Those with less experience should never be afraid to call on the experience of others when necessary. The experience of other people should be considered as one of the resources available to a troubleshooter.

In summary then, there are three factors to keep in mind it comes to troubleshooting. When confronted with a problem, technicians must make sure that they:
- have an adequate knowledge and understanding of the machine concerned,
- use common sense and a step-by-step approach, and
- draw on their own experience and that of others when required.

AIDS TO TROUBLESHOOTING

It is unlikely that problems will be solved quickly and effectively if the relevant information is not available. Therefore the first step in troubleshooting a problem to which there is no obvious solution is to collect all the necessary data to solve the problem. The data required falls into two categories and may be either background data (information regarding the function, design characteristics, maintenance instructions, etc. of the machine), or operational data (information regarding the running conditions at the time of breakdown). In both cases, the maintenance technician who is required to troubleshoot the problem must be aware of all the potential sources of such data.

Background data sources

Manufacturers' information

Most equipment items are provided with an operating and maintenance manual and a set of engineering drawings from which information can be extracted. Additional information can usually be provided by manufacturers on request.

Maintenance history

Most companies keep some sort of record describing the history of equipment items which can yield useful information regarding previous problems. Where no formal record is kept those who have been involved with the equipment may provide information from memory.

Systems drawings

Electrical, hydraulic and pneumatic schematics and control systems drawings are essential tools when troubleshooting systems problems. The ability to read and interpret such drawings is a vital skill that must be developed by those involved in problem solving.

Process drawings

There are various types of systems schematics that can be of great assistance. Process flow diagrams and piping and instrument diagrams (P.&I.D.s) provide useful information regarding the relationships and interconnections between all process equipment items. It is often important to have an understanding of the operation of related items and systems that may have contributed to the problem.

Troubleshooting charts

Manufacturers and technical literature sometimes provide troubleshooting charts designed to help identify the potential causes of common mechanical problems. Although these tend to be very general, they can be a useful starting point in the troubleshooting process.

Operational data sources

Operating records

If an operations log sheet or other form of record is kept, this may provide important information

regarding the operating condition of the machine prior to and at the time of failure. Continuous chart recorders are especially useful in helping to determine whether any sudden changes have occurred.

Observers' reports

It is often necessary to interview operating personnel who were in charge of the machine prior to or at the time of failure. Other personnel may also have been in the vicinity of the machine and may be able to provide data. The process of extracting information from others is not always easy and often requires particular communicating skills. These are discussed more fully in 'Communication' on page 126.

Test readings

If the machine is still operable, valuable data can be gained from readings taken by specially installed test equipment. Such test equipment may include multimeters, pressure gauges, temperature gauges, flowmeters, tachometers, dial indicators, etc., which should be installed in positions that yield the most useful information. It is not always necessary to install additional instrumentation to generate this data because use can often be made of in-line instruments.

Condition monitoring equipment

As described in Chapter 10, considerable advances have been made in recent years in the field of fault detection using various types of equipment that can be used to monitor equipment condition either on a continuous or spot-check basis. If this equipment is available it may be the source of invaluable data for the troubleshooter.

Metallurgical analysis

The examination of failed components using metallurgical techniques may generate important information. Microscopic examination of failed surfaces can help to determine the type of failure that has occurred and non-destructive testing methods such as radiographic and ultrasonic inspection, magnetic particle inspection and liquid penetrant inspection can yield information about material structure and condition that may help determine cause of failure.

INVESTIGATION GUIDELINES

Before undertaking the troubleshooting process bear in mind the following guidelines that are recommended to help avoid simple mistakes that can seriously interfere with finding a solution.

- Start the investigation as soon as possible after breakdown has occurred. Time may obscure or obliterate vital evidence.
- Ensure that no evidence is destroyed. Do not disturb or tidy up the scene of failure until a proper examination has been made.
- Collect important pieces of evidence for more detailed examination. Handle and pack them carefully so they are not further damaged accidentally. Accurately identify the parts and components collected.
- Do not be too narrow in the investigation. Check out the surroundings and environmental conditions and approach the point of failure gradually. The cause of failure can often be remote from the point of failure itself.
- Avoid guesswork and drawing easy conclusions unless they are rigorously checked out. Remember that a cause has been established not when it becomes obvious, but when all other possibilities have been eliminated. Be sceptical and always cross-check vital evidence, especially that collected in the form of statements and opinions. People's judgments and perceptions are all fallible and can be subject to subconscious prejudices. The instinct for self-preservation in particular can significantly affect people's ability to be objective.

COMMUNICATION

During the troubleshooting process much of the data will be provided by other people in the organisation, particularly operating personnel, and consequently the technician needs to be skilful at extracting and interpreting information. A technician must first of all know where to go for information and, secondly, how to get it. An awareness of the functions and responsibilities of others in the organisation, particularly those involved with the machine in question, is important as is being on good terms with others and having their confidence.

Very often the key to making sure that the necessary information is forthcoming lies in **asking the right question**. The ability to ask the right questions will depend to a certain extent on a sound knowledge of the equipment and its mode of operation and function. This knowledge will greatly assist the questioner in knowing what to look for.

Information collected should be checked and confirmed wherever possible. Incorrect data can easily be supplied, either deliberately or inadvertently, and it is always wise to cross-check.

The troubleshooter must be prepared to probe thoroughly for what operating personnel may consider irrelevant information. The slight variation in operating procedures that may be adopted by different personnel may well have contributed to the problem unbeknownst to the operator. If the problem is an unusual one without apparent precedent, questions should lead towards establishing what changes took place around the time of breakdown and what was done differently that might have affected the equipment.

It must be remembered that communication is a two-way process. It is one thing to ask the right questions but it is another thing to correctly interpret the answers. Hence **listening** becomes a key element in the process. The first thing to be sure of is that the question has been understood. It is then important to make sure that the answer given actually answers the question asked. This will require careful listening and some cross-checking and confirmatory questions.

The importance of clear communication during the troubleshooting process cannot be over-emphasised. The solution to many a problem has been overlooked because of what somebody 'thought' was said.

SYSTEMATIC DIAGNOSIS

It is generally accepted that the process of troubleshooting can be systematised into a series of fundamental steps that are applicable to fault location for all types of machinery.

These basic steps must be worked through sequentially and can be described as follows:

Problem analysis

This step primarily involves collecting information about the fault so that the problem can be defined as accurately as possible. It is not anticipated that the machine should be dismantled at this stage and, in fact, it may still be in operation. All known information from data sources listed in 'Aids to troubleshooting' should be reviewed so that the nature of the breakdown can be described as accurately and precisely as possible. For instance, the term 'pump failure' is a very general description of a problem which does not give much of a lead in the process of troubleshooting. However 'pump shutdown due to excessive seal leakage' or 'pump shutdown due to bearing overheating' defines the problem in much more detail and gives a positive indication of the direction the troubleshooting process should take.

Preliminary inspection

Once the problem has been defined, a more detailed inspection of the equipment can be carried out. In particular, the general area in which the fault is most likely to occur should be investigated. In many cases, where the problem is relatively straightforward, the cause of breakdown may be immediately apparent in which case a repair can be immediately undertaken. If this is not the case, then a number of preliminary questions should be considered before the investigation proceeds further. Such questions include:

- Is there a fault-finding guide for the equipment?
- Have there been any changes or modifications to the machine recently?
- Has a similar fault occurred before?

If the answer to any of these questions is 'Yes', this opens up particular lines of inquiry which should be pursued before the process advances further. If the answer is 'No', then the investigation must move on to the next step.

Fault zone location

If the fault has not been located by this stage then the equipment should be mentally divided into functional zones which can each then be checked for operation. For example, if a petrol engine will not start and the problem is not obvious, it is normal practice to isolate the problem to either the fuel delivery system or the ignition system by checking each separately.

The key to locating the fault zone is to check inputs and outputs rather than to examine the zone itself. For example, if it is necessary to check fuel supply to an engine it is better to disconnect the fuel lead and check fuel flow rather than dismantle the carburettor. The troubleshooter must establish key points in the system where tests can be made to eliminate those zones where operation is unaffected by the fault until, finally, the fault is isolated to one particular part of the machine or system.

Zone investigation

Once the fault has been traced to a particular zone or system then a more thorough investigation can begin. In the case of a single machine element such as a carburettor, the components may now need to be dismantled and examined. If the fault has merely been isolated to a particular circuit or system, then input and output checks may again be required to pinpoint the particular element in the circuit where the fault lies. If it is not possible to make test measurements then it may be possible to eliminate individual elements by bypassing them or substituting components that are known to work.

The more elements that can be eliminated as operating correctly the simpler it becomes to find the faulty element.

Finding the cause

The purpose of troubleshooting, it should be remembered, is not just to locate the fault but also to find the cause. If this is not done then a repair may be made and the machine put back into service and a similar breakdown may recur within a very short time.

As a guide to identifying the cause of failure it is useful to recognise that the cause can be classified according to the manner in which the failure develops:

Wear-out failures Failures attributable to the normal processes of wear as expected when the component was designed.

Misuse failures Failures attributable to the application of stresses beyond the item's design capabilities.

Inherent-weakness failures Failures attributable to a lack of suitability in the design or construction of the component when subjected to stresses within its stated capabilities.

Recognition of the cause classification will be an important factor in determining what corrective action should be undertaken so that the real cause can be treated.

Replacement or repair

The decision of whether to replace or repair the faulty component may depend on the overall maintenance strategy of the organisation and the downtime involved. If a component is repaired then it should be workshop tested, if possible, before it is reinstalled.

If failure was identified as being due either to misuse or inherent weakness then the repair made must include action to avoid repetition of the failure from the same cause. This may involve changes or modifications to other parts of the machine or system, or may involve the selection of different materials.

Performance checks

Once the repair has been completed it is essential that the performance of the machine is checked to ensure that the fault has been eliminated and that the machine is functioning satisfactorily. Before finally returning the machine to service it is wise to ask the operator to test it to make sure that it is operating correctly.

The systematic approach described above can also be shown in the form of a flow chart. (Fig. 11-1)

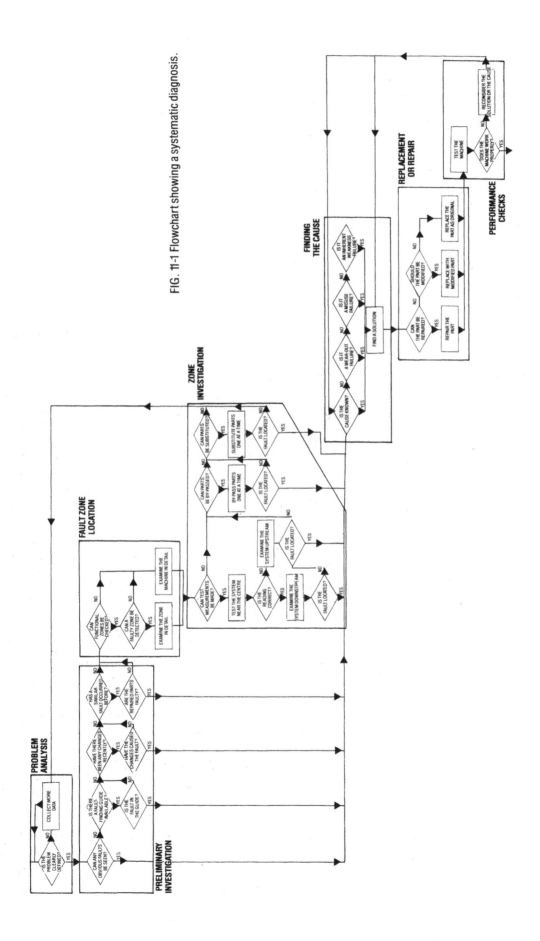

FIG. 11-1 Flowchart showing a systematic diagnosis.

ATTITUDES TOWARDS MAINTENANCE

As Robert M. Pirsig suggests in *Zen and the Art of Motorcycle Maintenance,* the manner in which maintenance work is approached can make all the difference between success and failure. It is often said that someone has the right temperament for a particular job and there is, in fact, a certain kind of temperament that is best suited to maintenance work. The point is that few people are born with the ideal temperament for any particular job but it is possible to learn to control responses and to develop those aspects that are best suited to the job.

Peace of mind

Maintenance work requires thought and attention. The work can often be very exacting, involving fine tolerances, complex assemblies and demanding problems to solve and it requires considerable concentration. Ideally the technician should become involved in the work to such an extent that other thoughts are excluded for the duration of the job.

The prerequisite for good concentration is peace of mind. The ideal state of mind in which concentration can develop is one that is uncluttered with stray thoughts, worries or anxieties, and able to focus solely on the problem at hand. It is rare to achieve sustained peace of mind but it is possible to learn to clear the mind for short periods and to push aside distracting thoughts and emotions. When actors take the stage they leave behind their worries and frustrations, and attempt to become the characters they are portraying. This requires a deliberate mental trick of emptying the conscious mind of the normal stream of thoughts and emotions and replacing them with the details and characteristics of the role to be played. Maintenance technicians should learn to play the same mental trick before undertaking a particular task. They must switch into the role of maintenance technician, empty their minds of all distractions and leave themselves free to concentrate on the task ahead. Peace of mind is a state in which the mind is uncluttered and receptive and free to apply itself to the task ahead.

The other important factor in achieving peace of mind is self-confidence. This was discussed in Chapter 11 with respect to troubleshooting and the same analysis applies to maintenance work as a whole. Self-confidence develops with a knowledge of the subject. Maintenance work has become more demanding as equipment has become more sophisticated and technicians must be prepared to continually upgrade their skills by doing research and asking questions. When a new machine is installed, borrow the maintenance manual and check it over, then when a job comes up you will feel more confident. Before starting any job about which there are doubts, time should be spent asking questions and checking through whatever technical information is available. Often the doubts can be overcome and the job then commenced with self-confidence and peace of mind.

It is almost impossible for anyone to concentrate and apply themselves properly to a task if some major catastrophe has just occurred in their personal lives. In this situation people are likely to make mistakes and hence be a danger to themselves and to others. Under these circumstances people should be treated as if they had a physical injury and put on light duties. In other words, they should be given tasks that are not mentally demanding until they recover. Mental injuries can be far more debilitating than physical ones but, regrettably, this point is often not recognised by employers.

Method

Just as the troubleshooting process demands a methodical approach, so does maintenance work in general. Complex and demanding tasks cannot be approached in a haphazard way if they are to be performed efficiently and without mistakes. There is always a shortest route between the beginning and end of a complex task and the first step that must be taken to find this path, in all cases, is to **stop and think**. This is the cornerstone on which a methodical approach is built and technicians should constantly remind themselves of the need to plan their work and to carefully consider the requirements of each step before proceeding.

For extremely complex tasks technicians can receive help from the Maintenance Planning Department who may go to the extent of producing a critical path network if it is warranted. In most cases, however, they must do their own planning and must try to anticipate the needs of the job in advance and equip themselves to cope with each phase of the job as it arises.

A few minutes spent at the beginning of a job thinking it through may save hours of work later.

Enthusiasm

Enthusiasm is the energy source that drives us along until the task is completed. There should be plenty of enthusiasm available at the beginning of a job but if it is an unpleasant one that does not generally arouse enthusiasm, then some sort of psyching-up process may be necessary. There are plenty of examples of other activities where this process is necessary, especially in sport, and it should not be considered as odd or inappropriate for maintenance technicians to psyche themselves up for a particularly unpleasant or difficult job any more than for an athlete to do so before a world record attempt.

The problem with enthusiasm is that it can easily start to leak away if things go wrong, and once it has dissipated, it is very difficult to replenish. The secret is to find a way of maintaining sufficient enthusiasm to carry through to the end of the job. The way this can best be done is to avoid, wherever possible, the situations that lead to loss of enthusiasm and the principle weapon in this struggle is the one described in the previous section – **method**.

The types of situations that can occur, but that can be avoided with good planning and a methodical approach, are such things as out-of-sequence assembly and unavailability or unsuitability of spare parts. Nothing can be more frustrating or have a more disastrous effect on enthusiasm than to reassemble an item and then find a washer or a spring left over and then have to disassemble and reassemble all over again. The key to avoiding this sort of situation is to adopt a methodical approach to the layout of components and to keep checking the assembly drawings during the process.

The same applies to the problem of spares. It is wise to personally check the availability of spares and correctness of part numbers before a job is started. This way it is known in advance what can and cannot be replaced and more care can then be taken to preserve items for which spares are not available.

The other things that can affect the supply of enthusiasm relate again to the mental state of the individual concerned. Impatience can cause severe problems and will frequently cause mistakes to be made. Thus impatience becomes self-defeating and rather than speeding things up can in fact slow things down to a standstill by undermining the supply of enthusiasm. Anxiety is also a mental state which can seriously interfere with performance and cause mistakes to be made, which in turn saps the supply of enthusiasm. Anxiety usually stems from a lack of confidence which causes people to be tentative and indecisive and thus liable to make mistakes.

The answer to avoiding these problems is to ensure that the necessary peace of mind is established before the job is started and to maintain this with self-control as the job proceeds.

If peace of mind can be maintained and a methodical approach adopted, then enthusiasm should last the distance.

Ego

One of the things that can seriously affect someone's ability to see things clearly is their own ego. It is easy to be convinced of one's own correctness and to ignore the advice of others or to ignore vital evidence that does not fit in with the opinion formed. This is a dangerous condition and can lead to mistakes.

It is vitally important that maintenance technicians remain objective and open-minded at all times. If they have a hunch about something they must be prepared to back it up with evidence and if the evidence does not support the theory, then they must be prepared to drop it rather than press on regardless.

Machines demand respect and do not bend to the will of human beings. Most of the time they are predictable but at other times they can be highly idiosyncratic (unpredictable). It is no good standing before the broken down machine and loudly declaring, 'But, it's just *got* to be the such and such!' when the evidence is to the contrary. In those impasse situations the technician must be prepared to recognise that it is he or she who is wrong, not the machine, and then continue to search for the vital evidence that has obviously been missed.

It is recommended that maintenance technicians show a little humility in their approach to their work; arrogance will certainly lead to trouble sooner or later.

The mechanic's feel

Finally, something needs to be said about what can be described as the mechanic's feel.

The properties of materials vary widely and not all can be treated in the same manner. Some materials are elastic and tough and can withstand high loads and stresses, others are brittle or soft and must be treated with great care. Technicians must appreciate different properties and handle different materials accordingly. They must be aware, for instance, when dealing with brass fittings that too much torque can easily strip a thread whereas high tensile steel fittings can withstand much higher stresses.

When dealing with threaded fastenings in general, they should appreciate the difference between **finger tight**, **snug** and **tight** and be able to interpret these three conditions in different situations.

The technician must also be aware of how to handle precision surfaces and those with fine tolerances. Both bearings and seals include components machined to high degrees of accuracy and it is vital that the technician treats such components with the respect needed to protect them.

In order to develop a feel for the materials with which they work, technicians must be aware and think about what they are doing. Like so many of the skills associated with maintenance, the mechanic's feel is not something with which the individual is born but a skill that can be learnt. It can only be learnt, however, if the individual cares enough to want to find out, and takes the trouble to observe, to question and to think about what he or she is doing.

SAFETY

Like maintenance, safe working practice is a preventative activity. In the same way that the objective of maintenance is to prevent breakdown, the objective of safe working practice is to prevent accident, injury and loss. Good maintenance technicians are as familiar with, and as proficient in, safe working practice as they are in good engineering practice. In fact, it would be more appropriate to say that safe working practice is an essential element of good engineering practice.

SAFETY-RELATED INCIDENTS

It is usual to use the word *accident* but there is really no such thing. When things go wrong there is almost always an identifiable cause and incidents are rarely, if ever, the result of some mysterious unseen hand. Safety-related incidents are the result of human error and failure to follow correct procedures and, therefore, can be avoided. They are not just 'accidents' over which no-one has control.

There are a number of ways in which incidents can be categorised, the most useful being according to the consequences.

- **Personal injury** Personal injuries resulting from incidents may range from minor cuts and abrasions that require little or no medical attention to serious disablement or death. Injuries that result in lost time are often recorded as indicators of safety performance.
- **Damage to plant and equipment** Incidents that result in minor damage to plant and equipment are often not recorded unless they cause lost production time. Serious incidents may threaten the local environment as well as causing major damage and loss of production.
- **Environmental damage** Incidents that lead to environmental damage are of increasing concern. Discharge of hazardous liquids and gases and disposal of solid waste are the most likely causes of damage.

Serious incidents may result in extensive damage to the environment, plant and equipment as well as injury to personnel.

Causes of safety-related incidents

Safety-related incidents may occur as the result of a number of factors. It is important that the significance of those factors for the performance of the maintenance function is clearly understood.

Error

Maintenance technicians, like other workers, make mistakes. Mistakes can be costly and result in serious injury and loss. There is always a reason why mistakes are made. Often it is due to carelessness or lack of concentration. It may also be due to lack of familiarity or lack of training. Maintenance technicians must take responsibility for their work, concentrate on the job in hand and, if in doubt, seek direction. The goal of every maintenance technician should be to eliminate error from their work.

Poor work practices

Maintenance technicians must follow correct work practices, some of which are generally applicable while others will be specific to the individual workplace. Part of good engineering practice is the application of sound work practices that reduce or eliminate the possibility of error, protect personnel and equipment, and create conditions in which the highest possible quality of work can be performed. Maintenance technicians must be familiar with, and follow, correct procedures.

Poor communication

Many safety-related incidents occur as a result of poor communication. It is the responsibility of maintenance technicians to ensure that they comply with workplace requirements by communicating verbally with those affected by their work and by following other procedures such as the completion of job sheets and reports, work permits and incident reports. Maintenance technicians are required to work closely with production and other personnel, and good communication is crucial to the effective performance of the maintenance function.

Poor management practices

Maintenance work is conducted within an organisational framework and climate that is established by management. It is management's responsibility to ensure that organisational structures, lines of communication, procedures and systems promote a safe working environment. Management is also responsible for the appointment of qualified staff at all levels and for providing adequate training.

Faulty equipment and materials

It is possible that safety-related incidents may occur as a result of faulty equipment or materials supplied to the enterprise. Such incidents can be guarded against by the institution of adequate testing and inspection procedures, particularly in relation to critical items of equipment and specialised process materials. The responsibility for the development of such procedures lies with management, but the implementation may be a maintenance responsibility.

Poor design

Poor process, system and equipment design can easily lead to unsafe working conditions and the potential for serious incidents to occur. Once again, the responsibility for ensuring that design standards are met lies with management. However, maintenance staff must play an important role by informing the design process and by providing a feedback link to ensure that design problems are identified and corrected in the light of plant experience.

By identifying the possible causes of safety-related incidents in this way it is possible to determine ways in which maintenance staff can contribute to the prevention of such incidents. It must be assumed that safety-related incidents will occur unless active measures are taken to prevent them. Maintenance technicians must, therefore, be proactive rather than reactive in their attitudes to safety, and strive at all times to adopt practices and to take measures that are designed to prevent safety-related incidents from occurring. The remainder of this chapter outlines the principal ways in which this can be achieved.

SAFE WORKING PRACTICE

Workplace behaviour

Strict codes of behaviour must be observed in the workplace in order to protect personnel and minimise the risk of damage to plant and equipment. Maintenance technicians may be required to work in any area of a production plant and, therefore, must exercise caution and pay particular attention to their behaviour. The following guidelines regarding general behaviour should be followed at all times:
* Never fool around in the workplace.
* Never run along corridors, on staircases or anywhere else in the plant.
* Do not play practical jokes.
* Do not play with fire, electricity, compressed air, gas or water systems.
* Never interfere with any machinery or equipment unless authorised.
* Never throw things around the workplace.
* Do not enter restricted areas.
* Never interfere with safety equipment.
* Never distract others from their work.
* Always follow regulations requiring the wearing of protective clothing or use of protective devices. Maintenance technicians should at all times:
* be thoroughly familiar with the work area;
* concentrate on the job in hand, but also be aware of what is happening around them; and
* observe correct procedures.

Maintenance technicians are the 'ambassadors' of the engineering department and their behaviour in a production plant should be exemplary.

Protective clothing and devices

One of the most fundamental aspects of working safely is wearing the correct protective clothing and using appropriate protective devices. Maintenance technicians must be particularly conscientious in this regard because, by being required to work in differing situations, they are exposed to a wide range of conditions and hazards.

The following guidelines should be observed by maintenance technicians at all times.
* Wear plain, tough, close-fitting overalls and keep them buttoned up. Loose or flapping clothes may get caught in rotating machinery.
* Wear any special protective clothing supplied by the employer.
* Wear safety shoes or boots and keep them in good repair.
* Keep long hair under a tight-fitting cap or net.
* Do not wear jewellery on the job.
* Wear personal protective equipment appropriate to the job in hand.

The following protective equipment should be worn when necessary:

Head protection
- Safety helmet

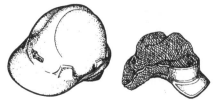

Eye protection
- Safety glasses or goggles

Ear protection
- Ear muffs or plugs

Hand protection
- Strong gloves

Respiratory protection
- Respirator or mask

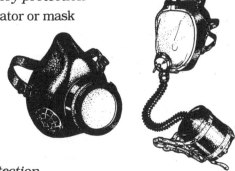

Foot protection
- Safety shoes or boots

Body protection
- Overalls

All protective clothing and safety devices used should conform to the relevant standards.

Housekeeping

Good housekeeping means keeping the workplace clean, ordered and tidy so that unnecessary hazards are eliminated. It is an important part of working safely and good engineering practice, and involves the following:
- Keeping work areas and benches clean and tidy and free from discarded material.
- Keeping floor areas clean and free from spills of oils or other liquids.
- Returning tools and equipment to their proper places of storage.
- Returning unused materials to the appropriate storage area.
- Disposing of waste and refuse in appropriate receptacles.
- Keeping aisles, accessways and exits clear and free from obstructions.
- Ensuring that all signs and notices can be seen and easily read.
- Ensuring that all necessary safety equipment is kept in good condition and is available for use when required.

Use of hand tools

Good maintenance technicians *never* blame their tools! Tools must be kept in good condition at all times and the right tool for the job must be used.

Using an incorrect, improvised or defective tool can lead to safety-related incidents and consequent injury. The following guidelines should be observed:
- Use the correct type and size of tool for the job.
- Check the condition of any tool before it is used.
- Do not use tools that are worn or damaged.
- Keep cutting tools sharp and protect them when not in use.
- Never use a tool for any other than its intended purpose.
- Store and carry tools safely.

Use of portable electric power tools

Electric power tools should be treated with the same respect as hand tools. They must be kept in good condition at all times and used only in the manner intended. Technicians should observe the following guidelines:
- Inspect electric power tools carefully before use. Check the casing, brush caps, switch, lead and plug to ensure that they are not broken or damaged.
- Do not use a tool that overheats and smells of burning.
- Do not carry or suspend a portable tool by its lead.
- When using portable power tools, hold them firmly and securely. Be prepared for the high torque of large capacity tools when they begin cutting or if they jam. Do not put them down until the tool head has stopped rotating.
- When necessary, hold the workpiece firmly in a clamp or vice.
- Do not use tools when they are wet or let tool leads drag through water.
- Protect other workers from sparks, hot metal etc. by using screens or barriers when necessary.
- Protect, cover or signpost tool leads to avoid tripping hazards to others.
- Always use appropriate personal protection.
- Do not leave unattended tools plugged in.

Use of compressed air and compressed air tools

Compressed air can be dangerous if used for anything other than its intended purpose. Maintenance technicians should observe the following guidelines:
- Check that air lines and fittings are in good condition before they are pressurised.
- Hold the end of a hose to stop it whipping when the air is turned on.
- Never use compressed air to clean clothes or any part of the body.

- Never blow down a bench or machine tool with compressed air. Metal filings and chips may be blown 10 metres away.
- Do not allow air hoses to become kinked.
- Follow safe working procedures similar to those for electric power tools for holding, starting, working with, putting down and disconnecting compressed air tools.

Use of explosive powered tools

Explosive powered tools are devices that use an explosive charge (the cartridge) to fire a projectile (the fastener) into hard materials. They can also be very dangerous weapons if used incorrectly. The following guidelines should be observed:
- Observe any warning notice attached to the tool and follow instructions carefully.
- Always use protective shields provided.
- Always use protective eye and ear equipment.
- Protect other workers by displaying a warning notice and using screens and barriers when necessary.
- If a tool fails to fire when triggered, wait for five seconds before removing it from the work surface and unloading it in a safe manner.
- Never attempt to close the breech, or cock an explosive powered tool, by 'crashing' it closed with the hand covering the open end of the barrel.

Machine guards

Maintenance technicians are frequently required to remove, replace and maintain machine guards. Guards are vital for the protection of all workers and maintenance technicians should take a particular interest in their condition and effectiveness. If a technician considers that the condition or design of a machine guard makes it ineffective and constitutes a potential safety hazard, the matter should be attended to or reported to the supervisor.

Danger tags and lock-out

There are several methods by which maintenance technicians are protected against the inadvertent starting of machinery or systems when maintenance work is being carried out.

Danger tags are used to indicate that certain switches or valves must not be operated. The technician working on a machine or system must attach a danger tag to the control switch or valve that supplies power or other input. The name of the technician, date and time should be written on the tag. Only the technician who attached the tag and signed it may remove it. When more than one technician is working on a machine or system they should each attach and remove their own tags.

When a danger tag is considered insufficient

protection, equipment should be locked out or isolated by more positive means. Electric circuits can be isolated at the switchboard and the main switch locked in the off position with a padlock. Similarly, valves can be padlocked in the off position. Danger tags are also used in conjunction with locks, and the responsibility for the attachment and removal of both lies with the technician working on the equipment.

To protect maintenance technicians and other workers from hazardous liquids and gases, spade blanks can be installed in pipelines. These should also be used in conjunction with tags to identify the technician concerned.

Every plant will have its own standards and procedures regarding tagging and lock-out of equipment. Maintenance technicians must follow correct procedures at all times.

Work permits

In order to gain approval to carry out work in a plant or plant area a maintenance technician will normally require a permit to work issued by the production department. This permit will ensure that all necessary safety procedures have been complied with and that it is safe for work to proceed. In some cases special permits to enter vessels or carry out specific types of work may be required in addition to a general permit.

The permit system is designed to protect all workers associated with an operation. It ensures that no work can commence until all necessary safety procedures have been complied with and that no plant or equipment can be restarted until all work is satisfactorily completed and there is no danger to maintenance or other workers.

Maintenance technicians must observe the permit system at all times and recognise that it exists for their safety.

Emergency procedures

Maintenance technicians may be working anywhere in a plant when an emergency occurs and so must be familiar with the emergency procedures that apply in each area. Evacuation procedures, assembly points, location of safety equipment, nature of local hazards, and location of accessways and exits should all be clearly understood.

Maintenance technicians should request special training sessions on emergency procedures if they are asked to work in an area that is unfamiliar to them or if they are in doubt about how to protect themselves and others.

HAZARDOUS MATERIALS

Maintenance technicians are likely to encounter many kinds of hazardous materials and it is vital that they are able to identify them and recognise their properties. An international code of symbols is used to indicate the general properties of substances in order that they can be handled appropriately.

Class 1: Explosives

These are substances or articles manufactured or used to produce a practical effect by explosion or a pyrotechnic effect, for example: gunpowder, gelignite, fireworks, fuses and detonators.

Class 2: Gases

Class 2.1: Flammable gases Flammable gases are those that ignite on contact with a source of ignition. Most flammable gases are heavier than air, and as such will flow to low areas, such as drains, pits, valleys, etc. Some gases have a subsidiary risk classification: poison (2.3) or corrosive (8), etc. Examples are acetylene (dissolved) and liquified petroleum gases (LPG).

Class 2.2: Non-flammable compressed gases Non-flammable compressed gases are those that within themselves are not flammable when exposed to a source of ignition. Some of these gases are liquified. Generally most non-flammable compressed gases are heavier than air, in some cases up to 6-7 times heavier. Some non-flammable gases can have a subsidiary risk category of oxidising (5.1) or corrosive (8). Examples are air, refrigerated liquid and oxygen (liquid).

Class 2.3: Poison gases Poisonous gases are gases which are liable to cause death or serious injury to human health if inhaled. Most poison gases have a perceptible irritating odour. Some of these gases can also have subsidiary risks such as being flammable (2.1), oxidising (5.1), corrosive (8) or can in some cases be both oxidising and corrosive. Generally most poison gases are much heavier than air. Examples are chlorine (gas), methyl bromide and nitric oxide.

Class 3: Flammable liquids

Flammable liquids ignite on contact with a source of ignition and have a flash point not higher than 61°C. Substances which have a flash point above 61°C are not considered to be dangerous by virtue of their lower fire hazard. The vapours from all substances of Class 3 have the property of a more or less narcotic effect, and prolonged inhalation may result in unconsciousness or even death. Examples are petrol, kerosene and paint thinners.

Class 4

Class 4.1: Flammable solids The substances in this class are solids possessing the properties of being easily ignited by external sources, such as sparks and flames, and of being readily combustible or of being liable to cause or contribute to fire through friction. Examples are sulphur, phosphorus and picric acid.

Class 4.2: Spontaneously combustible substances The substances in this class possess the common property of being liable spontaneously to heat and to ignite. Some of these substances are more liable to spontaneous ignition when wetted by water or in contact with moist air. Some may also give off toxic gases when they are involved in a fire. Examples are carbon, charcoal (non-activated) and carbon black.

Class 4.3: Dangerous when wet The substances in this class are either solids or liquids possessing the common property, when in contact with water, of evolving flammable gases. In some cases these gases are liable to spontaneous ignition due to the heat liberated by the reaction. Some of these substances also evolve toxic gases when in contact with moisture, water or acids. Calcium carbide is an example.

Class 5

Class 5.1: Oxidising substances These are substances which, although in themselves not necessarily combustible, may, either by yielding oxygen or by similar processes, increase the risk and intensity of fire in other materials with which they come into contact. Oxidisers may cause fire when brought into contact with finely divided combustible materials and may burn with almost explosive violence. Examples are calcium hypochlorite (swimming pool 'chlorine') and sodium peroxide.

Class 5.2: Organic peroxides These materials may be either liquids or solids. They support the burning of combustible materials. Under prolonged exposure to fire or heat, containers of these materials may explode. Many organic peroxides may react dangerously with other substances. Violent decomposition may be caused by traces of impurities such as acids. Decomposition of these substances may give rise to evolution of toxic and flammable gases. Examples are benzoyl peroxides and methyl ethyl ketone peroxide (MEKP).

Class 6: Poisons

Poisons are substances which are liable to cause death or serious injury to human health if swallowed, inhaled or by skin contact. They are divided into toxic substances (Class 6.1 (a)) and harmful substances (Class 6.1 (b)). These substances can be in solid or liquid form. Nearly all toxic substances evolve toxic gases when involved in a fire or when heated to decomposition. Examples are calcium cyanide and lead arsenate.

Class 7: Radioactive substances

This class includes materials or combinations of materials which spontaneously emit radiation. An example is uranium.

Class 8: Corrosives

These are substances which are solids or liquids possessing, in their original state, the common property of being able more or less severely to damage living tissue. Many substances are sufficiently volatile to evolve vapour irritating to the nose and eyes. A few of these substances may produce toxic gases when decomposed by very high temperatures. Also some substances in this class can be toxic. Poisoning may result if they are swallowed. Examples are hydrochloric acid and sodium hydroxide.

Index